Be sure and always
Up Stream from the

Cardboard Ponies *to* Buckin' Broncs

To John Maatta

Claudie Demarest

Claudie

Sweet Pine Hills Publishing Company
Stanwood, Washington

Cover Design by Kathy Campbell

Printed by Gorham Printing
Rochester, Washington

First edition, second printing

Printed in the United States of America

ISBN 0-9712502-0-0
Library of Congress Control Number: 2001093578

Dedication

To Auverne, my daughter
who encouraged me to write my stories
and produced this book

To Elsie, my wife
who was my loving partner for 62 years

To Douglas, my son
and partner in the operation of our ranch

To Sandra, Craig, Brian, Paula, Marty, and Crystal,
my grandchildren
who have brought much joy into my life

To Tiffany, Giovanna, Amanda, Reid and Rylan
who have made me a Great Grandpa,

And to Mabel, Helen and Lucille,
my sisters
with whom I share so many memories

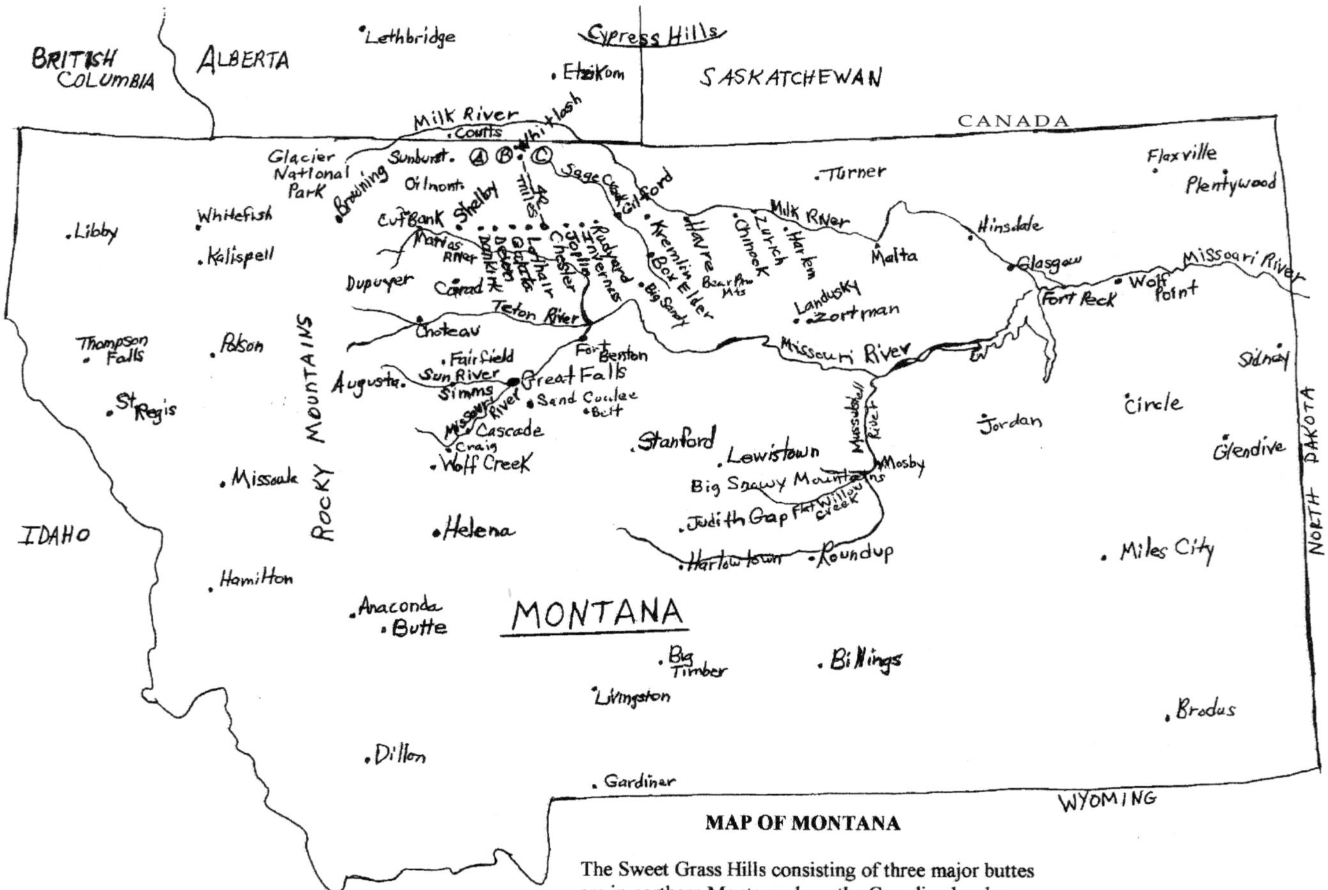

MAP OF MONTANA

The Sweet Grass Hills consisting of three major buttes are in northern Montana along the Canadian border. Whitlash is located between Middle Butte and East Butte. The buttes are labeled: A = West Butte, B = Middle Butte and C = East Butte.

Foreword

Claudie Demarest is an original Montana cowboy and cattle rancher. Today, in the year 2001, Claudie is approaching his 89th birthday and he's still riding horses and working the Demarest Ranch alongside his son Douglas. The Ranch was founded in the early 1900's when Claudie's father, Harry Demarest, homesteaded in Montana at the end of an era known as the "Open Range Cowboys." That era was the time period when thousands of cattle were being trailed into Montana from Texas, Oklahoma and California, the big cattle round ups were taking place on the open range, and Montana's colorful 'wild west' history was taking place.

Harry Demarest married Amelia Brown, the daughter of John Brown, a blacksmith and homesteader. John and Harry were among the original early settlers in this region of northern central Montana known as the Sweet Grass Hills. With the help of his only son, Harry built the Demarest Ranch from the original 160 acres. Today the Ranch is 7,000 acres.

The roots of the Montana cattle industry, and thus the cowboys who made it possible are traced back to Catholic priests, who in 1842 herded a few cattle into western Montana from Oregon. By 1846 this herd at St. Mary's Mission reported a head count of about forty cattle, which at that time was the largest herd in Montana. These first missionaries brought the cattle in hopes of teaching the western Flathead Indian to raise their own food. Father's Ravalli and DeSmet hoped this would dissuade the Flatheads from crossing the Rocky Mountains and war with other tribes competing for buffalo. Deer Lodge Valley with its lush grass became an area of major cattle ranches. A large demand for meat was a result of the gold rush of 1862 at Grasshopper Creek near

Bannack. This in turn created the first demand for cowboys in Montana to herd and care for the thousands of cattle being brought into the state.

As one reads Claudies' memoirs, you can't help but reflect upon how simple things were in his earlier days, yet how difficult simple ordinary tasks were by today's standards. Mail was delivered by horseback and supplies were delivered via jerkline, a string of six or more teams of horses pulling several wagons.

As a child I spent many summers helping out at the Ranch, going on cattle drives, branding, barrel racing horses during O'Mok'Sees at the Brown Ranch, or digging up arrowheads around the Indian rings. Today, there are still old wagons and even a well-preserved sheepherder's covered wagon in the barn on the ranch. Yes, the Ranch truly is a magical place. Grandpa's routine started with rising at dawn, going down to the barn to milk Swissy the cow, feeding the work horses their oats in the corral then getting ready for another 10 to 12 hours work. No such thing as "lunch" at the ranch, the noontime meal is "dinner" which is the biggest of the day with suppertime being the smaller evening meal. Times around the table were memorable; it seems visitors and neighbors would always drop by, right around dinnertime to enjoy a good visit with lots of knee slapping and laughing.

Claudie's memoirs are an original rare account of the end of those wilder western days. When reading his stories you can almost hear the old piano and fiddle playing in the background as you drift back to an earlier Montana, days now forgotten.

—Craig Douglas Dieffenbach

Contents

I / Harry and Amelia, My Parents

II / Early Memories

III / Horse Tales

IV / Elsie & Our Family

V / Punchin' Cows

VI / Ranch Work—Then and Now

VII / Fifty Years of Spaying

VIII / Community Life

IX / Crossed Paths Along the Way

Index

Introduction

The following stories are Dad's memoirs of his many interesting, funny and often very unusual experiences. He often tells us these stories, and I asked him to start writing them down, and write he did! The past few winters he spent writing everything he could think of, including explanations of how things were done in his youth and as a young man. I believe this is a wonderful record of our heritage. Progress has been so rapid in the past two generations, that his grandchildren and great grandchildren are now far removed from the way things were in the not so long ago past. While typing his handwritten stories into my computer, I find it amazing that when I was a child these long horseback rides and cattle drives of a minimum of forty miles or more were commonplace.

By Auverne Demarest Dieffenbach

Dad has been crazy about horses since he was very small. When he was a little boy, he got a pattern of a pony about four inches long. He traced them on cardboard, like the kind from the back of a tablet. He then colored them with crayons, branded them, and wrote their name on the back. He had a big collection of these cardboard ponies that he played with. He still has them today. When my brother Doug and I were kids, every once in awhile he would get the box down from a high shelf, and we would get to look them all over. Then they would go back in the box and up on the shelf.

He also played with stick horses. He carved his brand into the wood. Then he would string them out in pairs and had six or eight stick horses all tied together with string and hooked to his little red wagon. He would take a hold of the lead team and away they would go pulling the wagon. Some of the stick horses were saddle horses, and he told me, "Most of them bucked pretty high!"

Soon he was riding on a real horse along with his Dad. Then he started breaking horses, trailing cattle long distances, riding in rodeos, and continued doing much of his work on the ranch horseback. Thus came the name for his book. Dad has never lost his love of horses and at 88 years old still rides and punches cows.

Dad was raised and lived his whole life on the same ranch. The Demarest ranch is located in northern Montana in the Sweet Grass Hills about a mile from Whitlash and about five miles south of Alberta, Canada. It is forty miles north of the nearest railroad at Chester. The following excerpts have been taken from an article written years ago by my mother, Elsie Demarest who researched and collected data about the Sweet Grass Hills.

> "Standing like three giant sentinels guarding over the prairie that surrounds them, they have served as important land marks from the time of Lewis and Clark...... The Sweet Grass Hills are located along the International Boundary in central Montana. They cover a relatively small area in the northeastern part of Toole County, the northwestern corner of Liberty County and with a small part of their foot hills reaching into southern Alberta.
>
> They consist of three main igneous uplifts that are completely distinct from one another. Rising over 3000 feet above the surrounding prairie, they are visible for great distances.
>
> The western most mountain goes by the name of West Butte today but it has also been known as Mt. Jean. The highest of the three, it stands 6977 feet above sea level with the main peak to the south and with the lower peak trailing northward into Canada..... Middle Butte, also known as Gold Butte, is a very conical mountain and consists almost entirely of its central spire. It's foot hills are all much lower and are covered with grass. It was the gold discovered on the northern slopes of this mountain that really started the development

of the area, although at the time of the discovery, 1884, it was part of the Indian Reservation. East Butte consists of two main peaks a short distance apart and connected by a high saddle. Mt. Brown, the northern most of the two, has an elevation of 6959 feet above sea level. In contrast to it's somewhat pointed-top neighbor to the north, Mr. Royal has an almost perfectly round top and is forty feet lower….. Besides the three main buttes, there are several smaller buttes, Grassy Butte located south of Middle Butte,….. Haystack Butte south of the main part of East Butte…….. and Mt. Lebanon to the east of East Butte."

Uncle Claude on Ford, Dad on sawhorse,
Uncle Lu on Maxie and Claudie on stickhorse.
Note the angora chaps.

I / Harry and Amelia, My Parents

HARRY DEMAREST COMES WEST TO MONTANA

Harry Demarest, my Dad, was born in Illinois. The family moved to Nebraska where he grew up on a farm near Talmage and they raised pigs and corn. In 1899 Dad, his brother Claude and their friend Frank came to Billings, Montana on a freight train from Nebraska. They were given free transportation from Omaha to Billings as they came out on a "Railroad Contract". Their friend, Frank Kraft, had left Nebraska in 1887 and worked on ranches in Wyoming, and then went to Kalispell, Montana where he worked the winter of 1898 in the timber industry. Frank returned to Nebraska in 1899 and told his friends, Harry and Claude about the vast open ranges in Montana and Canada. Dad and Uncle Claude decided to go to Montana with him.

They arrived in Billings in the middle of the night and just outside of town they hopped off the train. The three friends managed to swipe their suitcases off the "hobo wagon" and hid out all night under a railroad bridge until the railroad "boss man" left town. The railroad had cops to see that those men who they had transported from the east to work for them didn't escape. After staying in Billings a couple of days Dad and Frank bought saddle horses and saddles and rode north. Dad got a job on Flat Willow Creek at a sheep ranch and Frank went a little farther north, also getting a job on a sheep ranch at Cat Creek. Claude got a job in Billings working with a carpenter. For the next two summers Dad also worked for the railroad driving teams and building railroad grades, and freighted with a jerkline team between Billings and

Roundup as well as doing some ranch work. He returned to Nebraska during the winters.

In the spring of 1901, both Uncle Claude and Dad returned to Billings. Frank had written to Harry that the Sweet Grass Hills country and Canada was the best country he had ever seen, and they should come up north. They picked up the two horses that Dad owned and had pastured near Billings and rode to the Conway and Kennedy Ranch on Flat Willow Creek north of Roundup where they worked for a grubstake. Dad bought a bronc for a packhorse from the Conway and Kennedy Ranch. I still have the Bill of Sale for that horse made out in longhand on ordinary lined paper from that rancher. Dad and Uncle Claude were making preparations for the trip north to join up with Frank Kraft.

In late summer they started on their journey. When crossing Flat Willow creek, the bronc got spooked and bucked their pack off into the water. They captured their bedroll and frying pan, but lost what grub they had. They of course didn't know too much about "The Diamond Hitch". Anyhow, they headed for the Sweet Grass Hills and Canada, a little gaunt, but a lot wiser on how to pack a bronc who was carrying their worldly possessions. Dad said if Uncle Claude hadn't have been a crack pistol shot getting a few rabbits and prairie chickens along the way, they would have gotten a little more gaunt.

One night they came to an old deserted horse barn. It was raining, so they tied their horses up in the back of the barn and rolled their bedroll out just inside the open door. During the night one horse got into a little ruckus. Dad jumped up and got him straightened out and to have a little fun he yelled to Claude, "Jump out the door, here comes a loose horse!" Uncle Claude made a big jump into the rain and into a big pile of manure up to his knees. Dad said it was a real mess and of course Uncle Claude had to clean up before Dad would think of letting him get back into bed. I told Dad that he was lucky to get any rabbit or sage hen after that trick, because Uncle Claude was the food provider on their trip with his trusty pistol.

When they got to Fort Benton, the news was out that President McKinley had been shot. They figured out that the tragedy happened the night they were camped out in the Snowy Mountains on their ride toward the Sweet Grass Hills and Alberta. That date was September 6, 1901.

One evening they were a little north of Chester about suppertime when they ran across a sheepherder's wagon. There was a terrible wind and they were thinking that they couldn't build a fire because of the wind. So they asked the herder if they could cook a meal on his stove. The herder wasn't too friendly at all and says, "You eat out there and I'll eat in here." So they did just that. Several years later Dad said he got even with the sheepherder. Dad was in Chester with a freight team and takes his bedroll into the livery barn to unroll it on the hay (his hotel accommodations) and here lay that ole sheepherder passed out on a nice soft pile of hay. Dad grabs him by the leg and pulls him out of his nice "nest" onto some hard ground and says, "You sleep over there and I'm going to sleep here".

When they got around Snake Butte south of where Hill Post Office located later, they met a rider and stopped to visit with him. He was the foreman on the Spencer Ranch on Milk River, and he hired both Dad and Uncle Claude. Their final destination was the Spencer Ranch northeast of the Sweet Grass Hills on the north side of Milk River in Alberta, Canada. They visited Frank for one day at the Strode ranch where he was working in the Sweet Grass Hills, and then went on across the Canadian border to work for the Spencer cow outfit. Uncle Claude's job was on the fall roundup, and Dad started freighting posts from Chester to Canada to be used as a drift fence to stop cattle from drifting out of the area. When winter of 1901 set in Uncle Claude went back to Nebraska to stay and Dad and Frank Kraft both worked for Strode and Dorsey in the Sweet Grass Hills. They worked around the Hills for a year then decided to go into Canada and locate. They went by saddle horses and packhorse, riding through the Cypress Hills to Maple Creek, to Fort Walsh and to Medicine Hat getting jobs now and then for eating money.

Dad and Frank filed on a homestead just east of the town of Etzikom, Alberta and lived in a dugout sod shack that next winter. When spring came, Dad went back into the Sweet Grass Hills and went to work on the Tom P. Strode Ranch east of Whitlash. The Stirrup  was the Strode brand. Frank stayed on the Canadian claim and did what had to be done to prove up. He also worked out on other ranches and they both put their savings into cows. They recorded two brands  and 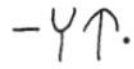. One was recorded in 1902 and the other in 1904. The brand office was Regina, North West Territories. I have these brand certificates framed and on my wall. This first location was referred to as "Spring Lake" and they bought 50 head of cattle. They decided they didn't like this location and in 1903 they sold it and moved a little further north to the Piegan Coulee on some land they purchased from A.W. Trussler. There was a log cabin on the place that had been built by Mr. Trussler.

They had 250 head of cattle when the hard winter of 1906 and 1907 blew in. There was deep snow lasting from October to May. They had fifty live cows when it was over. These cows had been following a bunch of range horses. The horses pawed the snow, so they all got a little groceries. There were a lot of Texas trail herds that also perished that winter. In those days some hay was put up for the saddle horses being used, but none for cattle. Dad said with all those thousands and thousands of cattle that perished, all that was left of them was a hide and bones and willow sticks, the willow sticks being the remains of what they had been eating. There was not enough flesh to make a stink or fill up a hungry coyote. Dad told me that he didn't have much equity left in this ranch, so he turned his share over to Frank Kraft. Frank married in 1909 and he and his wife Pearl lived in the log cabin for many years. They operated that ranch until 1948 when they retired to Medicine Hat. The ranch was sold to Fred Kreil. In the late 1960's a rodeo arena was built on the Kreil ranch.

Dad returned to the Sweet Grass Hills and worked for some time on the Strode Ranch and started over again. He bought two homesteads from Harry Larson and Frank Larson and moved in three shacks and put them together for his new home in the Sweet Grass Hills. He also bought 160 acres from Lila Herbert, the daughter of Edgar Simmons.

Harry Demarest on Diamond at Strode Ranch

Uncle Claude, Dad and Frank Kraft—1962

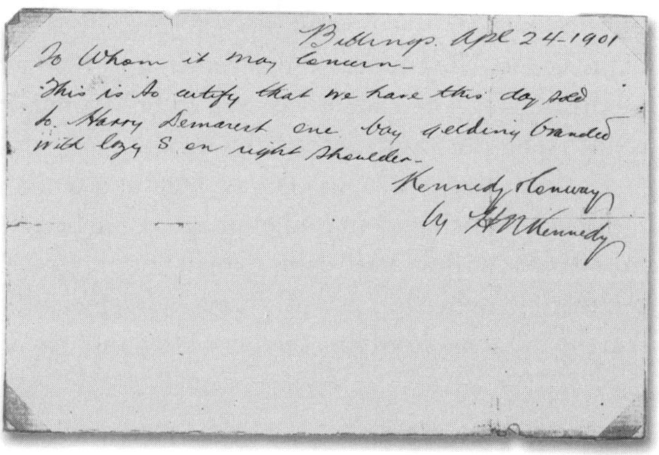

Billings. Apl 24-1901

To Whom it may Concern–

This is to certify that we have this day sold to Harry Demarest one bay gelding branded with lazy S on right shoulder–

Kennedy & Conway
by H M Kennedy

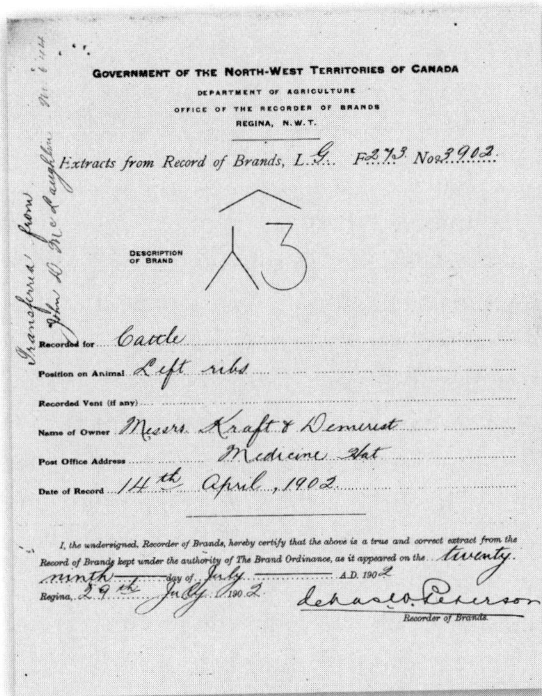

Transferred from John D. McLaughlin No. 422

Government of the North-West Territories of Canada

Department of Agriculture

Office of the Recorder of Brands

Regina, N.W.T.

Extracts from Record of Brands, L. G. F. 273 No. 3902.

Description of Brand

Recorded for Cattle

Position on Animal Left ribs

Recorded Vent (if any) ....

Name of Owner Messrs. Kraft & Demerest

Post Office Address Medicine Hat

Date of Record 14th April, 1902

I, the undersigned, Recorder of Brands, hereby certify that the above is a true and correct extract from the Record of Brands kept under the authority of The Brand Ordinance, as it appeared on the twenty-ninth *day of* July *A.D. 1902*

Regina, 29th July *1902*

Chas. W. Peterson

Recorder of Brands.

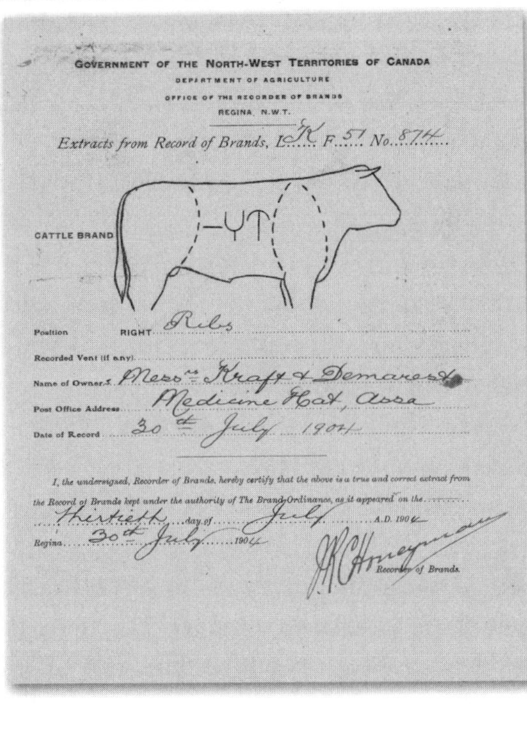

Government of the North-West Territories of Canada

Department of Agriculture

Office of the Recorder of Brands

Regina, N.W.T.

Extracts from Record of Brands, L. K F. 51 No. 874

Cattle Brand

Position Right Ribs

Recorded Vent (if any) ....

Name of Owners Messrs. Kraft & Demarest

Post Office Address Medicine Hat, Assa

Date of Record 30th July 1904

I, the undersigned, Recorder of Brands, hereby certify that the above is a true and correct extract from the Record of Brands kept under the authority of The Brand Ordinance, as it appeared on the thirtieth *day of* July *A.D. 1904*

Regina 30th July *1904*

J. R. C. Honeyman

Recorder of Brands.

BLIZZARD, JUNE RAIN AND TWO DEAD SHEEPHERDERS

My Dad Harry, tells about camping for the night on a hilltop about ten miles north of Chester. He had six horses hooked to a load of lumber. He fed them their oats and hobbled them, as there was lots of grass for them. He ate some chow and crawled into his camp bed which consisted of heavy blankets with a tarp for a cover. The next morning he woke up and a terrible blizzard was sweeping the country. He stayed in bed three days and nights and the morning of the fourth day the storm had quit and the sun was out. I have heard him say, "I sure needed a rest, but not for that long!" His grub box and the food in it was covered up with snow. He said if he had only taken a loaf of bread to bed with him, he could have eaten on that. He did have a box of crabapples that he had put under the covers with him to keep them from freezing. It was the 13th of October and the nights were getting pretty cold and frosty. The year was sometime before 1911 because that is when he got married, and at this time he was still single. He said that he got so hungry he felt like eating a bunch of those crabapples, but on second thought, decided he better not. With a belly full of them, for sure Mother Nature would be calling for a trip somewhere some place, so he just stayed hungry.

His horses were just a little ways off under a steep cutbank, protected some from the storm. Dad always carried a saddle and a bridle in his outfit and one of the workhorses was broke to ride. This was his "Little Ace in the Hole" in case he broke down and had to go someplace for repairs or help. He saddled up and headed for the Prescott Ranch about ten miles away, to get some food and rest. On the way he started finding dead sheep here and there covered with snow. When sheep get in a blizzard their noses ice up and they smother to death. Some were alive, but their eyes iced over and they couldn't see. One of these mounds of snow looked different than the ones covering the dead sheep. Under this pile of snow he found the sheepherder, who had perished in the blizzard. Dad went on up to the Prescott Ranch and told them. They asked him why he had not gone up to a very nearby cabin on the ranch, and Dad replied, "Well, I'm alive and maybe if I'd left my bed I'd be dead like all the others out there". He helped them for several days getting together the sheep that were still able to make tracks for themselves.

Another time Dad was in Galata for a load of lumber, also sometime before 1911. The month was June and word came into Galata that about four miles south of town someone had found a sheepherder who had chilled to death in this cold June rain. So Dad hitched four horses to his freight wagon. It took that much horsepower to pull an empty wagon in this gumbo mud, and he hauled the dead sheepherder to town. They put him in a side room of the saloon and sent a telegram to the owner of the sheep. The outfit who owned the sheep was called the J. B. Long Co., which was around Augusta. Dad and the others searched the corpse for an identification, of which they found none. But they did find some small change, enough to have a round of drinks at the bar. "And here is to you" to this unfortunate man who had lost his life in a June rain. Dad helped dig a grave and these cowboys made the coffin. They buried him on the northeast edge of Galata at the head of a small coulee.

About 30 years ago Dad and Grant Flage, who now owns this area where the herder was buried, looked for the grave, but after all those years the Montana prairie had hidden it well.

Harry Demarest and freight team at Whitlash

DAD MARRIES AND EXPANDS THE RANCH

On November 8, 1911, Harry Demarest married a local girl, Amelia Leone Brown, the daughter of Mary and John Brown. She was not quite eighteen years old, and he the bachelor, was nearly thirty-two years old. Their first home was the previously described place on the homesteads bought from the Larson brothers.

The wedding at Amelia's home at the Brown Ranch was to be at six o'clock in the evening. It was a very cold stormy day, and Mom's brother went to Chester to get the preacher. An old school chum of Dad's from Havre, Dr. Wilke (a dentist) was invited to the wedding and he came up with my uncle and the preacher. Well, the preacher froze his ears on the ride up because he only wore a hat, not a cap. My Dad, Harry, had quite a sense of humor and played a little joke on my Mom that day. He sent his sheepherder horseback to the Brown Ranch, three miles away, with a note to deliver to Amelia. In the note all he said that was that he was sorry, but he would be there at six o'clock that evening if he could get all the sheep in by that time. That's all they heard from him all day. He did arrive in time for the wedding! Being so stormy the wedding was only attended by the family. Sometime later they were surprised with a shivaree at their little home where they danced in the kitchen until daylight. Mom said that she kept that note from Dad for many years, until it finally was worn out and fell apart from us kids reading it and showing it to people so often.

The first year of their marriage Dad built up his small band of sheep and got some cattle. He built a chicken house, and pens for pigs they bought. In the spring they planted a garden. He also hauled a lot of groceries and freight from Shelby with a 12-horse jerkline string of horses for the Whitlash and Gold Butte general stores. Dad's sister, Helen, lived with them that first summer, and she and Mom did a lot of sewing. Mom painted and fixed up the little house.

During their early married years, Dad and a neighbor operated a threshing rig in the fall. They thrashed grain for neighbors with a horsepower machine. Ten horses in pairs went around in a circle. Inside the circle was a bunch of moving gears in a big steel box and from them was a rod that turned lying flat on the ground going to the separator which ran the belts and gears to separate the grain from the straw. When they moved the separator from one ranch to another it was equipped with wheels like a wagon, and a tongue so horses could pull it. Dad told me that when he went up to Alf Hicks' place to thrash going down that steep sideling hill, Alf came out with a saddle horse and tied his rope up high to keep the machine from tipping over as the ground was frozen and icy. Every time I go to the Hicks' Place today, I think of what Dad told me.

In August of 1912 Dad hooked his team to his spring wagon (or buggy) and took Aunt Helen and Mom 40 miles to Galata. Aunt Helen returned to Nebraska on the train, and Mom went to Great Falls on the train to wait for my arrival. On September 4th, 1912, Dad went "uptown" (Whitlash) to get his mail and found out he was a daddy and had a new baby son. Several weeks later Dad met the train in Devon. It was his first time to see me. We had a forty-mile team and buggy trip home to the Sweet Grass Hills. I have a letter that my Mother wrote to Dad from Great Falls before coming home. She told him that the hospital room and board bill for fourteen days was $30 and the Doctor bill was $25.

When I was a little kid, I thought there were only two "4ths" in the year. They were both "red–letter" days. The most important was September 4th, because this was my birthday, and the other was The 4th of July when we had a community picnic and festivities. Later when I went to school, I caught on that every month had a 4th. But they weren't as important as July 4th and September 4th!

In 1919 Dad bought his father in law's cattle and horses, and also the brand. Amelia's parents moved back to Great Falls, and later to Seattle, Washington. Her brother, Elmer, stayed on at the Brown Ranch.

Aunt Helen, Dad's sister also filed on a homestead and in 1918 turned it over to Dad. They moved from their original three-room home on the Larson Place when Dad bought the Frank Lowe homestead on Breed Creek in 1914. The two-story part of the house was the Lowe's home. Except for a few dormers and additions on to the west side, and some inside remodeling, it is the same now as when they acquired the property. Dad filed on a homestead in 1912 northwest of the main ranch, close to the Sutton ranch. Uncle Claude bought it from Dad and later he sold it to Robert Sutton.

Four hundred and forty more acres were bought from Edgar Simmons in 1915. This property joined Dad's land on the east side. Nettie and Lester Stott had a 360 acre ranch on Breed Creek that was up above the Hi Smith ranch. Dad owned the land where Whitlash is today and also other acres on each side of the road going from Whitlash east and he traded it for the Stott ranch on January 1st, 1917. This was strictly a trade and no money was involved. On Breed Creek the Stotts were milking cows and selling cream and butter to the creamery in Old Whitlash. They had to quit their dairy business because the wolves were eating up their milk cows and this was their livelihood. A. C. Strode, a cousin of T. P. Strode owned the land where the Old Whitlash townsite was. The Old Whitlash townsite had a big two-story hotel and a large mercantile store that was operated by Ellis and Mack who after leaving Whitlash had a store in Augusta. There was also a blacksmith shop, a saloon, a post office, and a livery barn. More land was added to the Demarest Ranch in 1918 above the Stott land up higher on the mountain. Another 360 acres were purchased from Ed Thompson and his wife Rosa who had a big family and a two story small pole house chinked with mud and grass. Rosa was a sister to Lester Stott.

The Stotts started out with a grocery store and the post office in a tar paper shack at the new Whitlash townsite (where it is now). Mr. Stott built the large two-story building that remains there today, calling their grocery store "The Ranchers Cash Store". It really should have been called "Jot-um Down Store", as the ranchers bought on credit. Mr. Stott was a great carpenter, and also used cement and stucco, the results of which can still be seen today in that original large building he constructed in Whitlash.

In 1927 Dad purchased the Alf Hicks homestead and ranch from the Stockman National Bank of Fort Benton. This piece of land lies north over the divide from Breed Creek. Dad also bought 80 acres on Breed Creek that was part of the Sid Griggs homestead. Sid Griggs was a brother to Mrs. Alf Hicks. His homestead shack was located in the northeast corner of his property, pretty close to the Hicks building.

In 1967 we purchased the Lila Herbert 160-acre homestead joining the north side of the Simmons place. Prior to 1967 we leased the land. Lila was a daughter of Ed Simmons and at this time they had moved to Eastern Canada.

Mrs. McBratney, a friend of the family who lived in Great Falls, homesteaded a small parcel of land for Dad in 1918. She lived in a little cabin up "McBratney coulee", and us kids, pulling our little red wagon, would go visit her about a half mile from our home, and of course there were always some goodies. This part of the ranch is still referred to as the "McBratney coulee".

In 1943 we traded for 160 acres that joined us on the north side of the Herbert field with Alvin Roberts who at the time lived on the Townsend Ranch. The land we traded, 214 plus acres, we had purchased from the two Manis kids that had inherited the land when the Manis Estate was settled. It is located east of the Simmons place, taking in the landmark "The Manis Ridge". The transaction squared up his property and ours too. The Manis brand was a R3.

In 1946 we purchased 2000 acres of the Arthur Fey Ranch on the east and south slope of Middle Butte in Toole County at $6.50 per acre. Arthur Fey had purchased it from Lynn Star. These 2000 acres takes in a big spring on Sheep Creek, on the original Henry Weideman 160-acre homestead. We still call it the Weideman spring. It is ice cold, coming from under a big sandstone rock. Henry was a relative of Norma Weideman Wehr Woods. Norma and Jimmy Woods have a ranch on Miner's Coulee and are our very close friends. Elsie and I stood up with Jimmy and Norma when they were married. Norma's first husband who died in an accident was Melvin Wehr, Elsie's first cousin.

In 1951 we purchased the Charles Murray Ranch of 681 acres at $25 an acre, it joins the Arthur Fey property we had bought on the east side of Middle Butte. The water rights on the Murray land were filed in 1894. A spring by the homestead cabin puts out 35 gallons a minute, never changes and keeps Trail Creek running water for miles. The 2681 acres is well watered with two live creeks, seven springs and seven reservoirs.

The present Demarest Ranch includes all of these acquisitions and is about 7000 acres. In the hills or rough country this would amount to more. This is where my three sisters and I grew up. There have been two homes on it since the time Elsie and I got married. Our two children, Auverne and Douglas grew up here and Doug raised his three children, Paula, Marty and Crystal here. That makes four generations of Demarests who grew up on this ranch. Doug has lived in Dad's and Mom's house for many years now.

Auverne and Tim live in Stanwood, Washington. Their children are Sandra, Craig and Brian.

Ed Thompson house, Steve Babokus by birdhouse —1935

Dad and Mom's first home

Harry and Amelia Demarest

Uncle Claude and Claudie—1913

SWAPPIN' HORSES

My Dad Harry, his brother Claude and a neighbor kid by the name of Cook decided to take a horseback trip through Kansas and into Oklahoma. They all lived near Talmage, Nebraska. Dad said that he rode a blind horse all the way into Oklahoma. According to Dad it wasn't uncommon to see a blind horse in their hot climate. The three rode onto an Indian Reservation in Oklahoma. My Dad, not being too happy with his transportation, traded for another horse for their return home. It turned out this horse wouldn't make a step unless it was led. Dad learned later that the horse he got in the trade was a Squaw's horse, which was always led by the Buck. So Dad rode all the way to Oklahoma on a blind horse and rode all the way back on that Squaw horse that had to be led. My guess is that the parties in this horse swap didn't go into too much detail about their horses! Each one thought that for sure he was getting the best trade. Dad said that the next time he would be certain that he was trading for a Buck's horse!

"Swappin' hosses is alright, if the other feller's got the best hoss!" I was given a little wooden plaque with this verse years ago and it hung in our dining room. I guess Elsie wanted to remind me not to repeat my Dad's mistake!

FANCY SURVEYING EQUIPMENT

Dad tells a story on himself and Edgar Simmons. They were just starting to survey a ditch to take water from the Simmons Coulee. Their equipment was a sixteen foot 2x4 with a leveling bubble fastened in the center and a leg nailed on each end. One leg was about two inches shorter, and by just keeping the bubble at the level mark the ditch would be surveyed at the correct down hill angle so that the water would run. They marked the route for the ditch with every move of the 2x4. Anyhow, when they stopped for dinner they happened to be right behind the barn at the Simmon's Place. When starting out again, they picked up their fancy survey equipment with the short leg in front and the long leg behind. They went ahead and surveyed and marked it out for well over a mile. Then they plowed the ground around side hills and little coulees to where they finally got to where they had stopped with their fancy surveying equipment, and then realized what a sad, sad, sad, dumb mistake they had made. They had been surveying making an uphill grade instead of a downhill grade! It no doubt maybe took a couple of weeks of hard hand labor to get that far for man and horses. Today with a cat and a dozer, it would still be a big job to make a ditch of that size.

Those Old Timers were a tough breed of cat and they went right back and started a new ditch. No doubt they broke a few more single-trees and harness-tugs, and also used more oats and hay for fuel. Most ranchers used a V-shaped plank with a wood floor and put a lot of rock in it for weight. Then they went through the plowed up ground, pushing the dirt to the down–hill side, until the ditch was made. This ditch wasn't one furrow and shallow like they made to prove up on homesteads. When I was a little guy, I remember asking my Dad why were there two ditches all the way through the field and as they went along, became further apart. Yet today, when we get a big snowmelt in the spring, we use that ditch to put water in some alfalfa ground and flood some of the native sod.

CHASED BY AN ELEPHANT

Dad and Murray Johnson were in Chester and there was a circus in town. It so happened that the elephant in the show got loose and people were running everywhere to get out of it's way. Dad and Murray were headed for cover when Dad glanced back and saw the elephant had turned around and was headed in the other direction. Dad stopped running, but being the practical joker that he was, he shouted, "Run, Murray, run!" Murray ran as fast as he could and slid and crawled under the depot, roughing up a little hide and clothing. Dad says if the ole' elephant hadn't turned back, he is pretty sure they might not have made it. He says he later heard that the same elephant had killed his trainer and had to be put down.

THE COW WITH PORCUPINE QUILLS

When I was a little guy, my Dad, and I were up at the Hicks Place. A cow came out of a muddy manure straw roof shed and she had porcupine quills in her nose. Dad builds him a loop to catch her by the hind feet to stretch her out and the loop missed her feet, but somehow caught her by the tail, which had a big mud ball on the end of it and the rope couldn't come off. So finally he dallied up and his shirt and jacket somehow got into his dallies and got to slipping. Pretty soon it had him bent over and the dallies was pinching his belly. I guess about that time he was hoping her tail would break off, but it didn't! So out came his jack-knife and he cut the twisted up shirt and coat away from next to his belly-button. Then her tail did break off next to the mud. He thought it would be a small hole in his clothes. Oh no! His whole front was naked! When we rode home that evening, there was a big doings of some kind at Liberty Hall. So we stopped out of sight over the hill and he put his coat on backwards and sneaked by the people out front of the Hall. I wondered why he didn't just cut the rope. His answer was that he had another shirt and coat, but not another good lariat. The cow? Dad roped her again, choked her down and pulled out the quills with some fencing pliers. At my age, myself and Duke was about as helpless as a screen door in a submarine to help out in this wreck.

BOOTLEG WHISKEY

When I was a teenager, there was a lot of bootleg whiskey coming out of Canada to the United States. Numerous loaded cars heading south came by our ranch. There is a big hill close to our place and after a rain a lot of these cars were getting stuck in the gumbo mud. These bootlegger's cars were those long bodied touring cars equipped with side curtains. There were no pickups in those days, so the largest and most powerful cars of the time were used but they were no match for that gumbo! Dad was kept pretty busy with his team of horses, a chain and doubletree pulling those outfits up the hill. Just north of home The Law had a little shoot-out with a loaded car of whiskey and shot out a couple of tires and the outfit went through a wire fence and down over the hill into a coulee. It tipped over and dozens of bottles of "giggle-juice" rolled through the grass. That was one car that didn't quite get to the muddy hill and on down the road to Great Falls. Too bad I wasn't old enough to have a party, 'cause I found bottles of liquor down in that coulee!

Dad once in awhile took a bottle for his good deed. He always liked to keep a little whiskey around for snake bites and a little gargle for sore throats. Sometimes he was given a cartwheel; however, he didn't expect any pay. This route went through Galata and across the Marias River to Great Falls and is still today called "The Boot-Legger Road".

DROUGHT IN 1918-1919 AND MINNESOTA HAY

I was only seven, but can remember real well the winter of 1918 and 1919, and of course having listened to my parents, Harry and Amelia, talking about it for years. The Sweet Grass Hills had been in a drought for two years and none of the ranchers had any hay or grass. The bankers told the ranchers they would lend them money to buy hay, and not to sell their cattle as World War One just being over, the cattle would be hard to replace. Dad said that cattle were really a good price that fall. Only one rancher in the Hills, who was financially better off and didn't need to borrow, or listen to his banker shipped his cattle to Chicago. And at that time in our area they had to go all the way to Minnesota to buy hay. Dad went back to Red Lake Falls, Minnesota and bought hay in the stack and a stationary baler to bail it and a truck to haul it to the railway station. He had it shipped to Galata, Montana. Dad had a man hired, John Stratton, to haul it forty miles to the Ranch with four horses hitched to a sled. We were lucky to get good quality hay. Some ranchers hired sharp-shod teams and mowing machines in Minnesota to cut the 'slew grass' that was sticking above the ice, then baled it and shipped it home.

Mom was left on the ranch with us kids, and the part I remember was we had to keep the ice cut on the creek so the cattle could get water. John Stratton who worked for us was on the road most of the time hauling that hay from Galata. I know he fed the cattle hay when he was there, but no doubt Mom had to do it when he was gone.

Dad shipped back the Ford truck that he bought to haul the bales to the railroad cars, and also the stationery baler was shipped home too. It was, as I remember, a long frame from the two ends. Dad traded it to a neighbor, who never baled any hay with it, for some horses. I think he must have used the metal for something else. I have often wished it had been kept.

Bill Schafer and Murray Johnson, our neighbors, shipped their bands of sheep to a reservation in South Dakota where there was feed. Bill went to take care of them, and Mary Schafer went along. She worked as a government seamstress at an Indian School on the reservation that winter. When spring came, they sold their sheep to a market back there; rather then have the expense to ship them back home to Montana. Anyhow, it didn't only bend, but broke every rancher in the Sweet Grass Hills. And that fall the cattle weren't worth the hay and expense that had been spent to support them.

Years later there were still two bales of that Minnesota hay, in our hayloft in the barn. Dad would remark, "Anytime from now on if I think we should buy some hay, I'm going up in the hayloft and take a good look at those two Minnesota bales, and say 'no' to myself and climb back down". And years and years later a fellow working for us, who didn't know the long ago Minnesota hay story, fed the two bales to the horses. Dad just bucked off his hat when he found out! But ranchers and cowboys are a tough breed and went right on raising livestock and fighting the elements of nature, and we still are some eighty years later, in 2001.

OUR FIRST CAR

The first car that we had that I can barely remember was an Overland with a cloth top and equipped with curtains that you snapped together if the weather was either getting you wet and cold or blowing your hat off. Dad bought it from a dealer in Galata by the name of Rinke. But what I remember most is that for years afterward Dad saying, "Yes, I gave $800 for a brand new car and only drove it 800 miles and parked it for good. It kept breaking axles." He gave it to Ray Stott to get it off the ranch.

My Mother, Amelia, told the following story about that car. "It was an Overland. And oh, how we loved it even if it was a fake through and through. Every place we went we broke an axle. It was about 1916 because Helen was a tiny baby. I was just crazy to drive, but no sir, Harry said, 'No sir, not as long as we have a young baby', I was not going (to be allowed) to drive. We didn't have a garage so we had to run the car into the shed and keep it there and it was an awful place to get it out 'cause (you had to cross) a creek and a bridge. First time he was gone out I goes gets the car out all by myself, and got the baby in a basket on the back seat and away I went! That was it, he never said another word. Never! And I didn't have a bit of trouble nor he never said I couldn't drive that 1916 Baby Overland. And it seemed every place we went after we had it about a year, was a broken axle." This was verbatim from an oral interview of Amelia Demarest.

THE OLD RANCH HOUSE GETS MODERNIZED

Dad bought the two-story hotel in Old Whitlash and used a good share of the material to build on the west addition of my folk's house. The addition included a kitchen, bedroom, bathroom and entry. This was in 1929. A carpenter from Shelby did the work. We also put in natural gas for heating that year. We were on a meter tapped into the main line for around thirty years until we had gas wells of our own and then had free fuel.

Also in 1929 Lester Stott drilled us two flowing artesian water wells, so before winter "took hold" we had bathroom fixtures and running water in the taps, gas for heat, the new addition on the house and a full basement, quite a change in one year. Soon grass started covering the trail to the little out-house.

HAIRCUTS

Dad used to cut a lot of hair around the neighborhood and of course cut all of us kid's hair. He would get the person up on a stool in the kitchen or outside in the summer, and tie a big white cloth around their neck and over their shoulders. His clippers were the hand kind, as there was no electricity in those days. He used to say, "There are only three days difference in a bad haircut and a good one, anyhow."

I remember so well when the younger girls were wearing their hair bobbed, but their Moms still had long hair done up on their heads in different fashions. Anyhow, my Mother decided to bob her hair and Dad of course obligingly chopped it off for her. She sure looked different! She was a little concerned about what the rest of the Moms would have to say at the next Ladies Club Meeting. I guess there was a lot of, "Amelia, what in the world are you trying to do? Are you trying to make us all feel old?" I do remember that in a few weeks Babe Furnell had Dad cut her hair. From then on many of our neighbor ladies got "bobbed". When Dad cut little kids' hair he would say, "Now you must sit real still or I might cut your ear off, and then you would look real funny, wouldn't you?"

Art Gothberg cutting Harry's hair

ROOTCELLARS

One of the most important buildings on a ranch or farm was the root cellar. They were built back into a hillside, with several feet of dirt covering the roof, and an entryway usually with dirt steps leading down into the hole in the ground. You didn't need any heat, as it never froze in this underground house where your vegetables were stored. Carrots were usually packed in sand in a big box. The sand kept them real crisp as if they had just been taken out of the ground. Cabbage heads were pulled out of the ground, the leaves left on, and a string attached to each big root which was tied to a nail from the logs that held up the roof. The potatoes were dumped into a big bin after being dug from the garden in the fall. I know that every family in these Sweet Grass Hills had a root cellar. I have been on lots and lots of ranches all over the country and just look around and there it is. Today there are other methods used, but lots of cellars are still in use.

Granaries were also used in those days as a place for food preservation. We "cured" our own hog hams and shoulders. These kept real good, buried several feet down into a bin of oats in the granary. The meat was covered with a light cloth material. They kept just as well in the summer as they did in the winter.

Remembering how we used to buy in large quantities, reminds me of a syrup story. My mother had just taken the lid off a one-gallon can of Karo syrup. It was sitting on the edge of the kitchen table. My little sister, Lucille, came along and was just tall enough to reach up and tip the whole gallon right on top of her head. And Mother came a running and Lucille was hollering to me, "Claudie, come get me, come get me!" I guess she thought I could "save" her, from Mother's wrath or the syrup, I don't know which. What a mess!

Cold Montana weather also helped us store food. Winters we used to order quite a number of boxes of frozen fish from Minnesota and we just buried the box of fish in a snowbank. This was before we had deep freezers. When we wanted fish for dinner we dug them out. Didn't even need a hook to go fishing, just a shovel! Karon Mader told me that her grandmother used to hang out chickens on the clothesline in the winter. They froze solid, and when you wanted chicken for dinner, you just plucked one off the line! That's how they did it in Choteau.

We had eight or ten bum lambs that we were feeding on a bottle They all disappeared one day and we rode a lot of miles lookin' for them. We finally decided the coyotes had gotten a big feed of lambs.

This was late in the summer and one day someone walking by the root cellar thought they heard a lamb blat and went into the cellar to investigate. Sure enough, the lambs had gone down into the cellar and the door at the bottom had gotten pushed shut. They had been in there for seventeen days. The last year's left over potatoes had not been disposed of, and the lambs had eaten all the "sprouts" which can get eight to ten inches long and are real moist. I guess that took care of them for water, and the spuds were about eaten up. When they came out into the sunlight outside, they started falling down, but soon they were of their feet again, gaunt, but alive and happy to get milk again.

MY FAMILY

I am the oldest of four children born to Harry Smith Demarest and Amelia Leone Brown Demarest. My three sisters and I were all born in Great Falls, Montana, and grew up on the Demarest Ranch in the Sweet Grass Hills. They named me Claude Smith Demarest, the Claude after my Uncle Claude, Dad's brother, and Smith because it was the maiden name of my Dad's mother. I was married to Elsie Rebecka Wehr from Butte, North Dakota, and we have two children, Auverne and Douglas. When we were first married we bought an old school house, moved it to the ranch and divided it into four rooms. At different times throughout the years we added on. It sits right across from my folk's house. I have continued to live on this same ranch where I grew up, and raised my family here. In earlier years I was in partnership with Dad, and now am partners with my son, Doug. He also raised his three children, Paula, Marty and Crystal on this ranch. My daughter, Auverne, married Tim Dieffenbach and their three children are Sandra, Craig and Brian. I also have three great granddaughters, Tiffany Orman, Giovanna Dieffenbach and Amanda Dieffenbach; and two great grandsons, Reid and Rylan Broadhurst.

Mabel Mae Demarest, "Chee Chee", is my oldest sister. Mabel used to help a lot with the ranch work and did all the raking at haying time. One fall Mabel drove the team and cleaned the manure out of a big sheep shed with me a hold of the "Johnson Bar"—it's the handle of a Fresno. There were all those shed posts, just twelve feet apart, and that takes a good "team skinner" to get around them without a "wreck". She also had to plow with a one-bottom blade to loosen the hard packed manure. We had one hell of a pile outside after about three day's work! She became a teacher and taught school for awhile. A neighbor fellow, Carl Iverson became her husband, and they ranched and farmed about eight miles north and east of our ranch. Carl started drilling oil and gas wells, and became a prominent figure in the Montana oil and gas business in addition to farming and ranching. Carl died in 1993. Mabel lives in Shelby and continues to manage their oil and gas business, going to her office every day. Their children are Ardelle, Dennis and Warren. Ardelle and her husband, Bruce Watkins own a beer distributing company in Great Falls. Dennis lives in Helena, and has been active in Montana legislation. Warren continues to operate the Iverson ranch near Whitlash, and also assists Mabel with the oil and gas business.

My next sister is Helen Amelia, "Honey" who married Wilford Weber from Flaxville, Montana. Helen also became a teacher, and continued to teach in Great Falls until she retired. Web died in 1974. She now lives in Kalispell. To Helen and Web were born: Steve, Byron, Jane, Greg, Mark and Stuart. Steve is an ophthalmologist in Kalispell. Byron teaches near Missoula and is also a naturalist. Jane is an x-ray technician and lives on a farm near Dutton. Greg lives in Alaska and is a CPA. Mark teaches in Twin Bridges. Stuart is an accomplished classical guitarist, writes his own music, performs in concerts, has put out several CD's, and lives near Bozeman.

Mary Lucille, "Chick", my youngest sister married Clarence Ritter from Rudyard. When Chick was young someone asked, "What are you going to be when you grow up?" She answered, "I don't know whether I'm going to be a School Teacher or a Lady". Three of the women in her life, her two sisters and Elsie were teachers, and our Mother was not a teacher, so she was a Lady. Well, I guess she decided to be a "Lady". She and Clarence had a farm in Turner, Montana for years where they raised their six children, Harriet, Donna, Russell, Carolyn, Ted, and Claude. Lucille and Clarence retired from the farm to Lakeside, Montana on Flathead Lake where Clarence died in 1995. Lucille still lives there. Harriet taught for awhile and now assists her husband in his Financial Planning business in Arizona. Donna is now a retired teacher living in Helena. Russell lives in the State of Washington and is a mechanic for a juice company. Carolyn is a "traveling" ultrasound technician, and works in many different hospitals on assignment across the country. Ted has his own scheduling engineering business, living in both Canada and the State of Washington, and travels a lot. Claude lives in Oregon and works as a Home Health provider. He also raises sheep and I hear one is named "CPR". Yep, it's the obvious. Claude Ritter saved one of his sheep by giving him cardio-pulmonary resuscitation! But Claude's sheep stories could be another book.

All of these nieces and nephews have families, and now some are even grandparents, so I've got many great nieces and nephews, and some "great-greats"! In their youth, they all spent time at the Demarest Ranch in the summers. The boys did ranch work and helped with the haying, and the girls helped Grandmother Amelia with work around her house and yard. Most all of them (boys and girls) have not forgotten helping my mother, their Grandmother, with the chicken butchering plucking and cleaning! That chore has been referred to as "the chicken genocide" by some! Amelia Demarest kept everyone busy!

My nieces and nephews continue to come to the ranch to visit Doug and I when they can, and we are always happy to see them. There is a lot of reminiscing about all the summers they spent on the ranch as kids.

DAD'S FAVORITE SAYINGS, EXPRESSIONS, & SONGS

Well, by the powers of the Great Horned Toad and a Cross-eyed Casaro!

Well I'll be tea-totally chawed up and spit out!

One of his tongue twisters he was always having kids try and say "real fast" was: *"The crow flew over the river with a chunk of raw liver in his mouth."*

Standing, Mabel, Claudie, Lucille; Sitting Helen, Harry, Amelia

Claudie, Mabel, Helen, Lucille and cousin Arlene Demarest

Brothers in law—Carl Iverson, Claudie, Wilfred Weber, Clarence Ritter

When something is wrong and you just don't feel good and didn't know what it was, *"Oh you must have the Cong-G-Go and can't Dis-per-crack-u-late."*

When he called anyone to get up out of bed, it was *"Change Cars!!"* This expression was from the railroad—in those days the conductor called it out to let you know it was time to change cars before the train split up and went two different ways.

First thing in the morning Dad would take a drink of water and say, *"This sure cuts the cob webs!"*

Dad called us kids *Siwasch* many times a day. I looked this word up, *"A word for an Indian Tribe"*.

Dad used to say if you got flour for people and oats for horses, you could winter in pretty good shape.

Dad had a few pretty good cuss words when he got mad. It used to make me kind of giggle.
"By the Bald Headed Jesus Christ Almighty"
"Oh Hell's Fire, Blood and Molasses!"

Whenever any of us kids lost a baby tooth, he always said, *"Don't put your tongue in the space where your tooth was, or a dog's tooth will grow there"*. We were pretty careful, but that's hard to do, and we had a big worry waiting to see what the new tooth was going to look like.

He would amaze little kids by doing the following, and then they would do it to him and also try it on themselves.
He would pull his nose up and his tongue would come out.
Then pull his left ear and his tongue went to the left side of his mouth.
Then pull his right ear and his tongue went to the right side of his mouth.

Then he would pull the skin under his chin down and his tongue went back into his mouth.

Little kids thought this was a great and exciting game!

When we had some good luck, especially looking for livestock, Dad always had a remark saying, *"We had Duck Soup!"*

A day when lots of work got done was, *"A large day"*.

When Dad found something he would say, *"There she is, the Dutchman said when he found his bull!"*

SONGS HE LIKED TO SING:

Old Dan Tucker was a fine old Man,
Washed his hands in a frying pan,
Combed his hair with a wagon wheel,
And died with a toothache in his heel.

And also the Nursery Rhyme, *"Sing A Song of Sixpence"*,

"Sing a song of sixpence, a pocket full of rye.
Four and twenty black birds baked in a pie.
When the pie was opened, the birds began to sing,
Now wasn't that a dainty dish to set before the king?"

II / Early Memories

FROSTY WINDOW PANES

Years ago, as a very young cowboy, not even yet a teenager, I remember how thick the frost would get on the inside of the windows in our house during real cold winter weather. The cold outside on a warm window glass caused it. Houses at that time weren't insulated, or as cozy and warm as today's homes. Storm windows or double panes as of today weren't thought of. And my guess with storm windows or double panes, they would have kept the cold air from the main window. I can remember how us kids would dig away the frost for a little peek-hole so we could see outside. I also remember what a mess it was when temperatures warmed up outside and this thick ice started to melt. It took all hands and the cook with towels to keep the water soaked up as the ice melted and the water ran down on the windowsills and the floor.

My Mother tells about one time I had got me a little peek-hole and could see a little of the outside world and told her that Dad just drove in with a team and sled to the barn. So she got me all bundled up to go out and come back to the house with him. Pretty soon when we didn't show up to the house, she went out to get me. She followed my tracks behind the sleigh tracks, going away from home. Dad had been leaving, not coming to the barn. Not too far away she found me, and I had lost my mittens and both hands were froze pretty stiff. She carried me, but every so often stopped and packed my hands in snow, and did this quite a few times before getting into the house. (What I have just written, I don't remember, this was as she told me in later years.) On getting into the house, I sure do remember, because it took both her and a schoolteacher who was boarding with my folks to stick my hands in coal oil to bring out the frost. I plainly remember that part 'cause it sure did hurt. She called up Dr. Melvin in Chester and told him what she had done. He said that having them partly thawed out before coming into a heated room was the right thing to have done before the coal oil treatment.

In those days everyone had a big supply of coal oil, because its what they used in the lamps for light, and it always set outside in the weather, and I guess was colder than snow.

I have never frozen my hands since that time, but they are the first part of my body to get cold yet today. I guess I shouldn't have worked so hard to make that little peek-hole in the frosted window pane, or else made it a little bigger to see what direction Dad was going!

In later years I have used snow to thaw out my frosted cheeks and nose. Afterwards the skin would always peel. This went on all winter with all the ranchers, 'cause you either were on horseback or on a sled behind a team of horses. There were no such inventions as hoods and parkas. We just had short coat collars. It was always best to have your face thawed out before coming into a warm room. Now days First Aid says not to use snow to thaw out frosted parts. Fortunately, I still have my nose and cheeks.

COLD WINTER MORNINGS

When we were kids growing up our house wasn't insulated and was a little chilly. In the living room was a little pot-bellied stove. At night a big chunk of coal was put in and banked with the draft partly closed so that the chunk of coal lasted till morning. A little metal grate in the ceiling let some heat into the upstairs. On real cold mornings the water bucket in the kitchen would be frozen with a good layer of ice, as the kitchen didn't have heat during the night. In the mornings, Dad got up first and fired up the kitchen stove, got the coffeepot on and some oatmeal a cookin, and also made bacon and eggs and hotcakes. He always had the plates and silver in the oven getting warmed up so they didn't freeze our hands or cool off breakfast. But that's the way it was and we didn't think anything different. Our water was carried from the creek in a bucket. When the creek was low we carried water for drinking from a spring in Stratton's field.

The ranchers were real saving on coal because it cost $8 a ton, a precious commodity, at the McDermott Brothers' coal mine at West Butte. Ruth Dolezal Burleigh's mother and grandmother were McDermotts. Very little coal was used in the summer months. Cooking was done with mostly wood. Before natural gas, oil or propane were available, everyone got in a good supply of coal for the winter months. The West Butte coal was real good coal, as it started easy with a few shavings or paper, and did not leave much ash. Some coal was hauled from Canada from just across Milk River. It cost less, but wasn't as good the West Butte coal. Coal-oil lamps were used for light. Natural gas wasn't available until the late 1920's.

HAYFEVER

When I was around five or six, I had hay fever real bad, and so did my Mother. Murray Johnson and his son, Vowers, had it bad, too. So one summer we all went back to Mount Clemas, Michigan for treatments. We took shots in the arm once a day and drank some terrible tasting medicine. I don't remember how long we were there. We boarded with a family and ate "family style" cookin', but I don't recall where we bedded down. But what I'll never forget were the house flies at the "Feed Trough". If you were a slow eater you just left the table a little gaunt. Maybe we didn't stay there long and found another place to eat, I don't remember. Something else I remember was that I had a little hat that I was pretty proud of. One day we were walking under some trees and a bird kind of fertilized my new hat! I wasn't too happy about that.

After coming home we kept taking that nasty tasting medicine. It did cure us because we both wound up with a dry nose, clear eyes and no sneezing quite a few years later.

My hay fever was so bad that I didn't start school when I was six years old. I did start when Mabel was six, two years later. In the summer I spent a lot of time in our cool dirt cellar which was next to the house.

MY FIRST HORSES

My first two horses were given to me. I was in my early grades and came home from school and here was a beautiful blood bay Shetland pony tied to the yard gate. His name was Duke. Dad had got him from a man at The Company C. Ranch. But he quite often bucked me off. This I don't remember, but Dad said one time when I got bucked off my nose was bleeding and my one arm seemed a bit long from pulling on the horn on that A-frame saddle. Dad says, "Well, I guess we will have to sell him", and I agreed with him. But the next day I told Dad, "If he bucks me off seventeen more times, than you can sell him."

Going back to Duke. A neighbor kid, Ellis Leach, had been giving me a bad time saying, "Bet I could ride him if he bucked!" One night after school he came down to stay all night with me. So he gets on Duke and lopes him up from the barn to the bottom of our hill and back and no buck-offs. I wasn't too happy with Duke. I say, "Ride him some more and really kick him good". Ellis 'took up a homestead' not far from the barn. That was sure one of my better days!

Claudie on Duke with Dad on Doodles

Some neighbors of ours were being chivareed. Aunt Vera was here, so she and Dad saddled up after dark and in the wintry weather rode seven miles to the Whoop-De-La get-together. Aunt Vera rode Duke. I think no doubt she left a few stirrup marks in the snow. Harry Flynn and Vesta at the Company C. Ranch were the newlyweds. Both mounts looked better the next morning

Frosty, my Christmas present

than the partygoers! Now days I guess if the heater in your car or pickup wasn't "up to snuff" you would just stay home and watch TV.

My next horse was a Christmas present. I was a teenager then. He was iron gray in color and his name was Frosty. He was gentle and a good-sized horse, and with my black hat and black angora chaps, I felt pretty tall in the saddle, you bet! I have a good pictures of both Duke and Frosty with me on top a trying to look like a cowboy.

1918 FLU EPIDEMIC

The flu epidemic in 1918 was a terrible disaster. It so happened that Mom and us kids were at Grandma Brown's house in Great Falls when we all came down with the flu. Dad found out that we were all real sick. He went to Galata and caught the train for Great Falls. We were terrible sick, but Dad did a good job getting us all back on our feet. A doctor never saw us, as the whole city was sick and they just couldn't make the calls. I was about six years old, and the part I remember is that our nourishment was Hershey Bars and grape juice, and Dad doing what he could to keep us all alive. He never got the flu. In later years I used to tell him it was his chewing tobacco that must have killed the germs. Harry Demarest's home remedies worked! We were lucky. Many families died in that flu epidemic all over the United States.

The same fall of 1918, mom and us kids were again at Grandma Brown's house in Great Falls when the armistice was signed and World War One was over. George Lauener and Art Gothberg, a Canadian neighbor, were at Grandma's house. They were on their way to be inducted into the service. We have a snapshot picture of them with us kids on the front porch.

CHRISTMAS ABOUT 1920

When I was about eight years old, I remember very plain when our family and the Elmer Brown family were all invited down to Forseth's home for Christmas Eve and Christmas Day. Sigurd, Torger, and Magnhild (Jeppesen) were Aunt Randi's brothers and sister. Their mother, "Grandma Forseth", (as we all knew her) had been widowed in Norway, and came to the United States with her four children and homesteaded just a little south of Grassy Butte. We went with Uncle Elmer's team of horses and triple box on a bobsled. There were Uncle Elmer, Aunt Randi, Mom and all us kids. They had the box tarped over and tied down. There were hay and blankets inside for us to sit on. But I do remember it was a real nice day. There was a little opening in the front of the box for the driver to stand, or for us to get out when we wanted to. It was an eighteen-mile trip from the Brown Ranch. I just don't know what year it was, but only the oldest of the Demarest and Brown kids would have experienced that trip. I remember when we got there; I walked up toward Grassy Butte and met Sigurd coming toward home herding his little band of sheep. My Dad didn't ride down with us, but waited to do his evening barn chores. That evening he came on horseback, and it was getting dark when he arrived.

The Forseth house was a little two-story home, which still stands today about a half-mile southeast of the end of the Strawberry Road. I will never forget how crowded we were getting bedded down for some sleep. I was never so glad to see daylight come in my whole life, 'cause they had me wedged in between Sigurd and Torger. Mom had warned them that I was a pretty restless sleeper and hell to kick. Years later after growing up, Torger would bring up that night with me sandwiched in-between. Boy, oh, boy!

Anyhow, the next morning Santa had sure enough been there. I really don't know how he kept from stepping on somebody making his Christmas Eve call. At about noon on Christmas Day we had a big dinner, and then headed north for home. Packages and boxes of presents made the trip down and lots of the same came back with us.

In those days, pretty near every family had what they called a foot-warmer made from a piece of metal about 18 inches long and maybe 6

Forseth homestead (2001)

inches deep with a space inside for a couple of bricks which looked something like chimney bricks. The bricks would be heated real hot in the cook-stove oven and slipped inside of this hollow foot warmer. I remember our foot warmer was covered on the outside with a tight weave green like floor covering material. A little door at one end opened at one end to put the hot bricks in. Anyhow, in that day and age, it kept a lot of tootsies warm by putting them on the old-time foot warmer. Our stage driver who brought the mail up from Chester with a team of horses kept one to use if he had a passenger on a trip.

CATTLE RUSSLIN' IN THE SWEET GRASS HILLS

One summer, I think 1922; Dad, Uncle Claude and myself mounted on my pony Duke, were riding in East Butte. We saw some cattle on Breed Creek in Meckler's field and Dad was curious, as they just didn't fit his cattle. We rode down off the high country to have a look. Well, it was a sight this young cowboy never forgot. These big three and four year old steers had been dehorned, their ears were cropped off, and it had probably been done just a few days before we came across them. In early days horns were always left on cattle and all ranchers earmarked with splits, crops or underbites for identification if the brand didn't show up. And these cattle all had worked-over brands. And me riding along and not missing a word of the conversations. My best friend in school was Mr. Meckler's son Frank. Listening to Dad and Uncle Claude talking it didn't take a detective to figure out who the real owners were under those worked-over brands. A few Strode  steers and a few Hi Smith  steers, and the balance were from the Webster Ranch on the Milk River in Canada that branded a 3CU. The 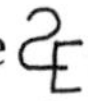 was changed into the  and the 3CU was made into the 800. The Strode brand had been blotched over. The Meckler brand was N-9. When I was a kid I was always looking at brands on livestock. I was really fascinated with brands and didn't outgrow it, I guess.

Several weeks later Dad and I rode out of the hills and went through Meckler's yard and met Pete walking up from the barn to the house. Dad says, "Pete, for an old cowboy you sure did a hell of a poor job branding them new cattle you got." Pete replied, "Ya, Harry, I know it, we branded them in an old rickety chute, should have headed and healed them, and stretched them out". That was all that was said, and then they went on talking about something else. This family was loved by every neighbor in the Sweet Grass Hills and was the life and fun of all community get togethers.

Later Dad had Tommy Strode (Mr. T. P. Strode's son) meet him up on Breed Creek and let him "Look–See" and satisfy for himself what was going on. Of course this ten-year-old kid on his pony was tagging right along and not missing a word. All these details on what was said stuck in my mind all my life, I never forgot, but as I'm writing this tonight, "yesterday I probably forgot in which room I left my hat". Tommy reported what he had seen to Johnny Morgan, our Sheriff of Liberty County. How Pete found out that he was reported I really don't know, but for a couple of days he hid out not too far from their home in an out of the way place with a bunch of cottonwood trees. Then he rode into Canada to the Cypress Hills and stayed with an old friend, for how long I don't know, but maybe just a few weeks.

I happened to be staying overnight with Frank at their ranch the night he rode back home from Canada. Frank and I had already hit the Sougans. Pete found out from their hired girl that night that the Law was holding his wife, Kate, so he put his horse away and went to bed. It's so clear in my mind the next morning how happy Frank was to see his Dad. After breakfast Pete walked over to the old time crank phone on the wall and called the Sheriff's office and says, "I'm home Johnny, come up and get me." I can still see and hear him tell the Sheriff that. Anyhow, he was "A Man" and wasn't going to let his wife "take the rap". Their hired girl was Edna Gothberg, a Canadian girl from just across the line, and we knew the family well.

My Dad said he noticed that summer at different times on riding through their ranch that most of the time Pete had Pretty (his horse) tied to a post and grained up and feeling his oats. At the auction sale of their property, Mr. Strode bought Pretty, a beautiful sorrel gelding. Dad bought a brand new "Weber" wagon with a high grain-type box. We still have it. I was at Meckler's staying over the weekend when several cowboys came to get the cattle. They were corralled and officially counted by our Sheriff and turned over to the cowboys to trail home, the cattle being minus a little hair, ears and horns.

All of the Meckler's dishes and private articles were packed into wooden barrels and stored in our hayloft until after he served his term in Canada. I think he served four out of eight years. Then Pete went to work in the Kevin Sunburst oilfield where Kate was already employed as a cook. Before Mrs. Meckler went to the Kevin Oilfield to work, she cooked for a drilling crew that was drilling the #1 Harold Brown gas well. Later they moved to another state.

Soon after this affair happened we bought the Hick's Place from the Fort Benton Bank. Years later we happened to find this big pile of horns in a large hole on top of a rocky cliff, not far from the Hick's Place buildings. Bob Thompson and his sister Dorothy Thompson came from their ranch at Jordan, Montana and took back the ranch that their Dad had sold to the Mecklers.

Years later, my folks and I stopped and visited with Pete and Kate not too long before they both passed away. They had always remained very good friends.

(Note—Some of the names in this story have been changed.)

TURKEY EGGS FOR MABEL AND A NEW HORSE FOR ME

Around the year of 1926 a neighbor of ours, Harry Flynn, who lived on the Company Ranch, had a little bunch of pet sheep that strayed away and went over to Hill, Montana. Mabel and I went over horseback and brought the sheep back for him. We wouldn't take the money he offered us, so he gave Mabel a setting of turkey eggs, and myself a two-year old horse. Little did we know at the time that the eggs and horse were of much more value than the cash. I was about fourteen and Mabel was about thirteen years old.

Those eggs kept us in holiday turkey meat for years. They were on their own all summer but when fall came that turkey hen would come home with her family who by then were near grown-up. Then we fed them grain. The night before we wanted a turkey to eat, I would reach up and grab one off the corral fence where they roosted at night. I would take him down under a tree next to the creek and hang him up by a wire tied to his feet, and take a sharp knife and stick it up his open mouth and bleed him. I know it sounds pretty cruel, but if you didn't use this method, those turkeys were just about impossible to get the feathers picked from them. Cutting the heads off like we did with chickens just didn't get the job done. After you killed a chicken they were put into a bucket of scalding water, taken out and then their feathers were easy to pluck.

The horse, whose name was Barney, was broke to ride. Later he was used as a workhorse for harness. He did a lot of hard labor on our ranch for years.

NO TIME FOR BOREDOM

When I was a kid I knew what I would be doing when I got home from school on Mondays. It was wash day, and my job was to help my Mother by cranking back and forth on the handle of the washing machine. No horseback rides for this cowboy on Mondays after school!

There wasn't such a thing as "being bored" for a teenager on a ranch, or even in town when I was a kid. I had lots of chores to do, choppin' wood for the woodbox, keeping the coal buckets filled and also taking out the ashes from the pot-bellied stove in the living room. I helped Dad with the milking, cranking the cream separator, taking the skim milk down to the trough in the pig pen, and taking milk to the "bucket calves" down by the barn. We planted a big vegetable garden and it was a lot of work to keep the weeds out in order that we would have plenty of veggies to eat for the summer and the next winter, too. There was always lots of work for all of us kids, but that was just part of everyday life on the ranch.

CHINAMAN WITH A RAZOR

One time when I was at a rodeo in Shelby, everyone was uptown lyin' and braggin' and whoopin' it up at Turk's Bar, and me old enough to get bucked off, but still not old enough to belly up to the bar for a short snort. So I was standing around back and noticed a Chinaman with long braids sitting in a chair next to the back wall. I kept lookin' at him 'cause he sure looked a little out of place, just sitting there kind of sleepy-like. A fight got started at the bar and that Chinaman really came unglued, jumped up and whipped out a razor and headed for the bar, waving it in the air.

This teenage cowboy was about the first one out on the street, but right behind me came a lots of company; cowboys, sheepherders, blondes, brunettes, and bar tenders. Never did I know if anyone got sliced up, but I knew for sure that this kid was still healthy. That razor looked plenty wicked and I'll bet it was what you call "Razor Sharp"!

MONTANA TO NEBRASKA IN THE RUMBLE SEAT

In the early fall of 1933 I went to Nebraska with my cousins, Harry Demarest Jr., and Rex Miller and another friend of theirs, in a coupe with a rumble seat. It was raining when we left, in fact, we were seven days getting to Lincoln, Nebraska, and it rained all the way. Two of us were always taking turns in the rumble seat, covering our heads with a blanket to protect us from the elements. Before we got as far as Great Falls, my cousin spots a combine setting out in a grain field. This is when I found out what an "Oklahoma Credit Card" was. It was a hose and a five-gallon can. They just drove up to the combine's gas tank and siphoned out gas to fill their tank. The tires on this car were nothing but "frog-skin" thick and the tubes were all patches. We patched tubes all the way. None of us had much money, and nobody wanted to spend any. In some town after dark in Nebraska, they drove around until they spotted a parked outfit that was just like what we were driving, that had a spare on the back. Harry gets out and he crawls under the parked outfit to get the spare. We drove around the block and got just about back to him when the owner came out of his house, got in the car and drove away, leaving Harry on his back with the spare tire in both hands! He got it undone just in time! So after several days with the Oklahoma Credit Card, and "borrowing" a few minor patches and glue, they picked up at stations, we did get there. Both my hands were in terrible shape from pumping up tires all the way from Montana with the old-fashioned tire pump.

I had my 21st birthday in Nebraska. I came back home in style in a Cadillac with Uncle Claude, Frank, Granddad Demarest and Cal Parmenter. We stopped at Connoleys Western Wear in Billings and Uncle Claude bought me a new black Stetson for my birthday. Velma Gardener always said, "You went to Nebraska as a boy and came back as a man!"

III / Horse Tales

MAXIE

Maxie was Dad's horse, but I used to ride him. I was a teenager, and had been to a circus in town where I saw trained horses prancing around and rearing up. I decided to train Maxie to rear up. He soon learned that when I raised my right arm up high in the air that was the signal for him to rear up on his hind legs. One day Dad was riding him while working some cows. Dad threw his right arm up and yelling "hi-yi, hi-yi" to move the cows along, and up reared Maxie! Dad was pretty startled and surprised. He didn't know he was punchin' cows with a "trick circus horse".

Claudie and Maxie showing off

ADVENTURES WITH SPEED, ONE TOUGH HORSE

Roy Todd was here one winter and we broke several horses. Speed was one of them who belonged to Johnny Dolezal. He had been a stud till he was eight years old. He was tough. A good cow horse was good at heeling calves. At the barn on leaving for a ride he stood good to get mounted, but out in the field he was hard to get on. He was tall and stout. He never did set me a-foot, but I knew that some day it would happen. I tied a ring on the end of a short soft rope. I would put the rope on his left front foot, pull his foot off the ground and take a couple dallies around the horn, step on, let his foot down and the ring made it drop off his foot. I had already ridden him about four years, and the last two years I guess we both had a better understanding of each other, at least on three legs.

Speed finally got a little stiff and we used him on the mowing machine for several years. Johnny did take him home the first year we started riding him. Anyhow, I bought him for $35.00. He was the toughest horse I ever rode and I camped on a lot of them.

This reminds me about the time I bought a new bridle. It was a split-ear with no throat-latch. I had it on Speed and was up in Louisa Anderson's field. It was hot and he was fighting flies, and off came my bridle. That's when the run-away started. I did manage to keep the bit in his mouth, but finally he got that spit out, then stuck his foot through the rest of the bridle and jerked the reins out of my hand. From there on Speed was in the driver's seat, runnin' and buckin', and I felt about as useless as tits on a boar pig, so I gets the old biscuit in one hand and a hand full of mane in the other. Several times he slowed down a little and I was tempted to quit him,

Claudie and Speed at the Canadian Border

but I was still healthy where I was and not having a parachute didn't chance it. Finally after around an hour he was getting hot, tired and winded and he let his passenger get off. I tied my belt around his neck and led him around the field until I found my new beat-up bridle. When I got home I real quick like got some leather and made a throat-latch. I have the bridle yet today, no more of them split-ear bridles without a throat-latch for this cowboy!

One time I was riding a bronc that was doing a pretty good job bucking and the short latigo broke and throwed me and my Cogshall on the rooftop of an old car parked in the barnyard. We rolled off the car landing on the other side in a pile. I still had both feet in my stirrups and a good hold of the horn, but no horse or buck rein.

LIBERTY COUNTY HORSE ROUND UP - 1929

In 1929 Liberty County had a horse roundup, gathering all the stray horses in the county. These were horses that were left by the homesteaders when they pulled up stakes and left for greener pastures. They were owned by banks and loan companies and accumulated into large numbers. These horses became a big nuisance by getting into people's property, and getting trapped in the fenced lanes in the Sweet Grass Hills area. Especially during summer, our lanes were full of the horses. They came to the Hills from down on the flats because some years the water holes and reservoirs down there would become dry. In the Hills we had live creeks that flowed across fenced lanes. Many Montana counties had these stray horse roundups.

Jimmy Morrison was our Round-Up Boss. The riders were Carlie Iverson, Ernie Price, Leonard Parsell, Swede Cady, Ralph Harris, Bob Robinson, Johnny Kossel, and myself. We had several different pastures in different areas of the County where we would put bunches to hold until we could trail them to the stockyards in Chester. We would corral them at night and two cowboys or "horse jinglers" would herd them out on grass during the day. When we finally got too big a number gathered to fit the stockyards at night, we put them in a field east of the Chester cemetery, but once a day they would be watered at the Great Northern Reservoir. Jim had a 1928 Chevrolet coupe and tried to keep us fed with sandwiches that he would get put up in town. Of course ranchers and farmers were glad to bed us down and feed us if we were there, but a lot of time we weren't near a ranch and our cookie wagon wasn't around, either. We did have a meal ticket at Tommy Smith's when we got to town, and we slept in an empty boxcar up at the yards, so you can guess the water taps and plumbing wasn't too convenient. We put in some tough hours with lots of dust mixed with it, but no dull moments!

At times we were a little short of riding horses, and we were always trying out the odd horse in the horse gather that had a few saddle tracks on his weathers. Ernie Price was a great bronc rider and after he pulled a few wet saddle blankets from these once broke horses, he would turn them over to Carlie and me. With those cactus plants all over the area, it made you take a deep seat in the cottonwood tree of your saddle and keep good track of the "biscuit", or "nubbin" or "Grandma"—all names that cowboys used to refer to the saddle horn.

A few of us did sneak away and took in the rodeo at Goldstone, Montana, up next to the border. Ernie took "first" in Saddle Bronc Riding and us two teenage cowboys got bucked off. Carl Iverson was the other teenager; he was about a year and a half younger than me. They had a stage built in the middle of the arena for the band that furnished the music. I remember real plain that the horse I drew was a real chute fighter and they had me belted down for a spell before they could let us out. The horse I drew did happen to be a strawberry roan in color; and he bucked me off up on the stage with the musicians. The band really welcomed me with some toe tappin' music. "The Strawberry Roan" would have been an appropriate tune. Anyhow, I was a bit embarrassed. I'm not sure if I put a spurred boot through the drum or not.

Ralph and I left the yards to go get a few horses southeast of Chester next to the Hill County line. Riding through town we stopped at the drugstore and he got a bottle of some kind of disinfectant as he had a bruised and skinned-up leg. We got out of town to a plank bridge across Cottonwood Creek and stopped on the bridge. He unbuckled his chaps and dropped them, took his Levi's off down to his spurs, removed the bandage from his leg and poured this medicine on the raw wound. He fainted and fell his full six feet, 'ker-plunk', and there he lay, stretched out and pretty much naked and both spurs jabbed in the bridge planks. I thought it pretty funny, but it must have been fifteen minutes before he came to and by that time I was getting a little worried. Then he had a big bump on his noggin in addition to the sore leg. A fellow came along in a Model A Ford, but we had the bridge kind of blocked. Anyhow Ralph went back to town, as he sure didn't feel too good. I went to where these few horses were and stayed with a bachelor all night who helped me bring the horses to town the next day. My helper was happy to get them out of his pastures. We did sneak this bunch out of Hill County.

Out north one day we had started to the yards and Jimmy's saddle horse played out. So we bunched the horses up, and Ernie throwed a loop, and on the end of it was a gray mare. We got her saddled while being "eared down". Jim took a deep seat in the cottonwood and we turned her loose. Of course we expected to see an explosion of horse flesh, but she just trotted off and started to whinner and a little colt came out of the bunch and run up to its Mommy and started getting a little lunch. I can still see Jim looking at what he couldn't believe, he was pretty happy, but a little embarrassed. Riding a mare with a colt following wasn't exactly the pride of a cowboy. We found out later that the gray mare had taught a whole family of kids to ride. The reason for her putting up such a fight is understandable, never having been roped and ruffed out before!

Another time Ernie's horse gave out and we did the same, but we had a real wildcat! Had to throw him to the ground, rolled him into the saddle, untied his feet and Price came up with him. On the trip to town, horse and man were in the herd and out. After a few miles he finally quit buckin' and we corralled them with the herd. Ernie didn't have to walk to town, and he took a lot of abuse, but he was a whoopin' and grinnin' all the way to Chester. The horse had a little less hair on his shoulders and was wearin' a real wet saddle blanket on his arrival into town.

We had a small bunch about 20 miles north of Chester with a gray horse among them that had given us the slip several times. He would just flat out-run us. So Jimmy gave them a big run with his 1928 disk wheel Chevie coupe, then we riders took over and made it to town. A few days later Louie Oswood asked Jim, "Do you have a gray hoss in your gather with a X-9 on the left jaw and full of notin' but life?" "Yes, we do Louie". After the chase he gave these horses, Jimmy's Chevie had four broken springs and would high-center on a road apple. It hung pretty low to the ground all the way into town and to the car hospital.

We had gathered around 3000 horses in the county, which covers an area of 24 miles wide and 56 miles long. A day was set for people to come to Chester and claim any of their own horses that they wanted, pay a round-up fee, and pay the county tax if it hadn't been paid. Or one could buy any horse they might want for a canner price. Elmer Brown bought two Blue Mouse colored mares; one a light blue, the other a darker blue color. I remember we picked them up northeast of Chester. The person that owned them wanted to get them out of his way as he had no use for them and he needed the grass for his milk cows. Grown horses were worth $5 per head; lightweight horses $3 per head as canners. The horses that weren't claimed or sold were loaded in stock cars and shipped to Butte, Montana to a cannery. Jimmy was paid for all the unclaimed horses that went to Butte. That was his pay and us Horse Jinglers were getting $6 a day and room and board when we caught up to it. Six dollars per day was big wages when ranch wages were two dollars per day at that time. Jimmy told us fellows we could pick out of the bunch any horse we wanted. I took a three-year-old unbranded Chestnut colored mare.

I remember we picked up a Bay gelding that had saddle tracks on his weathers and had been broke to ride (a little bronky though) that belonged to Howard Sailor on Sage Creek north of Gilford branded . He was a top horse, and we used him during the gather. The brand now belongs to Terry Stevenson and Mary Ann Stevenson, owners of the Sailor Ranch.

A lot of times in our gather you would see a mare with her sucking colt, along with her yearling, two year old and three year old, still staying together as a family.

1929 roundup, horses in Hawley lane

I was a seventeen-year-old teenager cowboy when I went on this horse round up. It was a great experience, one of many I will never forget.

GILFORD RODEO, 1930

It was 1930; I was eighteen years old and was at the Gilford rodeo to ride saddle broncs. I will never forget that at this rodeo a rider from Great Falls came out on a saddle bronc and the horse fell on him because the horse was flanked too tight and couldn't get up. At that day and age they used just one flank strap from the saddle, not a double flank like they do now. I was standing pretty close to the spill and took out my jackknife, cut the flank strap and up came the horse. But the rider was dead, having been crushed by the horse. It was a pretty scary experience for a young cowboy to see someone killed right before his eyes.

A little later the chute boss says to me, "Kid, your horse is in the chute; do you want him or shall I turn him out? Nobody will blame you after what has just happened". I just climbed up on the chute and got down on the horse. When they chapped me down on this ole bronc I'll admit the "S— was just a little bit above the sand." They flanked him real loose, and I made the whistle on a pretty easy bucker.

It was approaching dawn when I got home from that two-day rodeo in Gildford. About daylight I had just gotten my poor aching frame into bed, and Dad woke me up to tell me that Nettie Stott and family had been on a picnic up Breed Creek and she had not returned to camp. They had been searching for her all night. Dad told me to get up and saddle up and ride into East Butte and help with the search for her. After two days of saddle broncs and wining, dining, and dancing, I wished than that there had been another day of rodeo. Artie Morrison and I rode all over the west slope of East Butte, a lookin' for Mrs. Stott. We went up Ribbon Creek and up on the very south part of East Butte up by the Brown Eyed Queen mineshaft. The shaft hole at that time of the year was pretty full of snow. From there we climbed Mount Royal from the south and then off Mount Royal into the saddle next to Mount Brown. We went down Little Joe creek to Breed Creek and didn't find her. Anyhow, I don't know where she had been, but when daylight came she found her way back to their camp. I was sure glad when she showed up, 'cause I was still waitin' to catch up on a little sleep and do some "healing-up"after wining, dining, dancing and broncs!

THE THREE BEAUTIFUL DAUGHTERS AND THE BOX ELDER STUD

Mr. Strode had bought a stallion from a Mr. Culligan who lived in the town of Box Elder, a town south of Havre and north of Big Sandy. I left in the fall of 1932 to go get him. I rode Speed, a bay Hot Blood, and he was plenty tuff, no road was too long for him. We went by the way of Black Jack (on the north slope of East Butte) and the Bear's Den area.

I planned on staying at the Howard Sailor Ranch north of Gildford, Montana. His ranch was on Sage Creek, which starts on the east side of the Sweet Grass Hills and flows by Bob and Toni Dafoe's ranch, which at that time was the Harry Hoffner ranch. I had never been to the Sailor Ranch before, so pretty much kept Sage Creek in sight: It was I'd guess around fifty miles. At that time the homesteaders had most all pulled out for greener pastures, so very few people lived in the area. Fences were all pretty much fallen down, so cutting across country was easy going. I remember I never opened a gate all day, but I sure had to watch out for scattered barb wire so that my horse wouldn't get tangled up and cut on it.

I stayed all night at Sailor's and headed out for Box Elder. I went through the little town of Gildford, and I'm sure everybody's dog in town ran out to check me out. And Speed really wasn't too fond of dogs. He kicked a lot of puppy dogs; I never heard so much howling in all my life and I'm pretty certain they weren't all alive. So I never looked back. I knew quite a few people in Gildford as I had ridden in several rodeos there. Oh yes, and they were Howard Sailor horses that he furnished for the event branded a lazy .

I was happy to get out of town, and a little further on stopped at the Earl Rambo place and Earl said he would go with me. I was glad to have company and he knew the country to Box Elder better than me. It was all open country, lots of homestead shacks and no people were seen at all. We did stop at a dilapidated place where a family lived and asked the lady for a drink of water, which she got for us. But drifting out of the open door was some really beautiful piano music. I would like to have got a look at who was tickling the ivories, I'm sure she must have been beautiful, too. After leaving, Earl, he kind of squelched my happy thoughts, saying, "Bet it was her husband playing". Anyhow, that good drink of water and the beautiful music kind of made the miles a little more relaxing the rest of the way.

We got into Box Elder after dark and finally located the Culligans who owned the horse. We got our horses put away in a nice barn with hay and oats. This family had already had supper, and started preparing some chow for us. They welcomed us two cowboys like a couple of long gone sons who had decided to come home to the nest. We got sat down to a great meal. After forty or fifty miles a cold biscuit or a raw spud would have been good enough. Under the table was a trap door on hinges that went down some steep steps into the cellar. After we got set down, Mrs. Culligan decided we had to have some special goodies down under, so up we got and moved the table to one side and down the hole she went, and she fit the opening petty darn tight, which brought on a grin. And believe it or not, we moved the table two more times for more goodies, and there was no way we could talk her out of it. The hospitality was out of this world. Maybe she might have been thinking about a couple of potential "sons in laws"!

While this was going on, Mr. Culligan, a little man about 110# soaking wet, was a tuning up his fiddle. I wondered, "Now, what?" Their house was pretty small for a dance, but they did have Three Beautiful Daughters who were real easy on the eyes, made you not even want to blink. One daughter played the piano and one a guitar. Pretty soon the fiddle player went and got a gallon jug and it sure wasn't full of water. Things got pretty toe-tappin' lively. Soon the jug started getting' uncorked real, real often. The old guy could really jig and play his fiddle. The more he sipped on his jug, the faster his feet and bow were getting a workout. Quite a sight! And the one Beautiful Daughter who wasn't in the orchestra really liked to dance, so us two cowboys kept her happy and us a grinnin'! After a few hours, our fiddle player quit and staggered out and hit the hay.

They bedded Earl and I down on a cot in the kitchen and with all the giggle-juice, we were having a little trouble getting to sleep. If our fiddle player had been a little bigger man to hold more spirits, he could no doubt lasted till breakfast time. Pretty soon the Lady of the House, came by us with a flashlight and moaning and groaning, and kept saying, "Oh, mercy me, save me God, oh, mercy me, save me God!" and kept repeating it all the way through the kitchen. Our only guess was that she was in a big hurry to get to the little house outside. Sure hope she made it in time! Years later, anytime Earl and I met, he would repeat her word for word and laugh, and laugh, and laugh! The chow was delicious and the entertainment was "Whoop-De-La" and the night was a way, way too short!

The next morning it was snowing a little. The stud horse was a ways out of town, but luckily in the direction we would be headed. We knocked on this 'ole bachelor's door and went in. He knew someone was to get the stud. He was big and fat with bib-overalls on and really not too tidy or clean lookin'. He insisted we had to eat before that long ride to Hingham where Earl lived. He says, "All I got is a big kettle of boiled cabbage a-cookin'". Well, that sounded good to me 'cause I really did like plain boiled cabbage. About this time we could see and smell that this old guy couldn't control his bathhouse chores; he was pretty 'high' and those bib-overalls were wet from his belly button to the floor. And Holy Jesus! When he pulled the lid off that kettle and out came the steam and it mixed with this poor old feller's togs, it just wasn't too appetizing, would have about gagged a skunk, I'm sure! But Earl and I had to set there next to him and eat on a little two-by-four table; never dreamed one could hold down that much food and only take a couple of quick-like breaths. And you know folks; it was a good forty years later before I could tackle that menu again. Anyhow it seemed good to get out in the fresh air and clean white snowflakes. The old feller remarked as we left that he thought we should have eaten more on starting a forty to fifty mile ride. But we thanked him for his hospitality and headed west toward the Rocky Mountain front.

After going a ways, this stud refused to be led. He was real gentle-like, and I guess broke to work. We worked on his rear with a rope and all he would do is kick at our saddle horses, and no way could we drag him all those forty miles. Earl says, "Why don't you ride him?" "Yeah," I replied, "why don't you do just what you thought was a good idea?" "Tell you what", Earl says, "let's draw straws". Quick-like Earl jumped off and got a stem of white sage, and broke it into two pieces. I think yet today that both straws were the same length and C.S.D. was the one who had to lengthen out his short ladigo on the Cogshall saddle about three more holes; got it on the stud and took a deep seat in the cotton-wood. Well, he might have been broke to work, but sure as hell wasn't broke to ride! He felt as wide as a boxcar, and hit the ground pretty hard, but not too high and squealed like a pig getting cut, and all the while passing gas. I'm sure that old man had been feeding him boiled cabbage! After his little tantrum he lined out pretty good, but I'm getting a little more bow-legged every mile on his wide ole frame, so changed horses. Must have been midnight when we finally got to Rambo's. Boy oh boy, what a great fun night we had with the Culligans and their Three Beautiful Daughters; but what a hell of a day and half the night getting to Rambo's!

I left there the next day and went only as far the Ruth and Lester VanWetchel place north of Rudyard. Ruth VanWetchel is Elsie's cousin. The following day I made it to the T. P. Strode Ranch in the north side of the Sweet Grass Hills. Oh yes, we finally named this stud "Geronimo", but on the start of the trip we gave him a lot of handles, that just aren't printable.

THE BROWN MARE

Late in the fall of 1932 Edgar Woods and I went to a farm a couple of miles north of Hingham to get a stud horse for Squeak Johnson, a neighbor of ours at Whitlash. This was a stallion that had been raised by the Rambos, south of Hingham. A friend of mine gave me a nice looking brown mare that was with the horse that we were getting for Squeak. She wasn't' broke, so I roped her in an old barn and snubbed her up to my horse. When she lunged out the door, she kinked her neck, which made her nose about a foot from the ground. We went ahead and she followed along. About dusk we could see some farm buildings with a lot of trees, the Packer's place and it really stood out. At that time very few farms had trees. It was lightly snowing and the kids came home horseback from school. Dorothy Packer, who now is Mrs. John Duncan, was one of the school kids. We stayed there all night. Mr. Packer said, "You better leave that mare here; she will never make it to the Hills". We should have taken his advice.

The next day about dark, the mare gave out. One of the Colbry boys came along with a grain truck. They parked the truck in the ditch and we got her loaded. Edgar went with him up to the Dodds Keith Ranch that was about around eight or ten miles away. It was dark by then, and I followed up with the two horses. They unloaded her and we stayed there that night. Bud McDowell was working there and said he would call me when she got better. A couple of months later Bud called and said I could get her anytime. The next day he called back and said he happened to see her buckin' and playin' in the corral and she slipped and fell, killing her dead. That fall I guess just broke her neck.

Anyhow, a lot of different trip from the last one we made after the stud horse. No Pretty Girls, no fiddle, no jug, or no boiled cabbage.

A HORSE'S TAIL

When the Ed Gordons were on the Hawley Ranch, early one morning we corralled their cattle for calf branding. After eating breakfast I went down to the horse barn and my Stevie horse had his tail bobbed straight off, right next to the bone! I was so damned upset; it about made me sick. This was one cowboy whose saddle horse's tail- trim had to be perfect to my liking, right to a "point" and about six inches below his hawks. I knew right away it had to be Doug Gordon, Harley Umphrey and Gene Umphrey who pulled this awful trick on me. But what did I do to deserve this? Oh yes, and some shellac had been poured across the seat of the Hamley Committee saddle. This all happened in 1933, the year Elsie taught these kids when the schoolhouse was in Thompson's field. Then it came to me! Not long before I had ridden down from the Hicks Place and when I went by the schoolhouse, I noticed that these kids had saddled their horses at the last recess, ready for a fast go for home. I had stopped and unsaddled their ponies and went on my way. So I guess my little prank kind of backfired. Believe it or not, I turned Stevie loose for over a year till his once beautiful tail had grown out. I am still riding that Hamley and the shellac stain is still there. It's a good leather preserver, along with a few spur tracks on top!

Speaking of Harley and Gene reminds me of a few years earlier when they were not yet in school. I always stopped on my way home from riding in East Butte to visit the Umphreys when I rode through their ranch. Harley and Gene were little guys and I always wound up with one on each knee. On one occasion I had some canning jar rubbers for tie-downs on my spurs to my boot heels. Anyhow, several weeks later I stopped to see them, and both kids were running around with jar rubber bands around their shoes. Hazel, their mother, said they wore them every day since I had previously been there, a running and buckin'all over the place.

1934 CHESTER RODEO

In 1934 Carl Iverson, Jimmy Morrison, Ernie Price and I put on a four day rodeo in Chester on the lst , 2nd, 3rd and 4th of July. We cut some poles in our field in East Butte for three chutes, and had the plank gates made in town. These were what you called "shotgun chutes" with a gate at the end and on the side, and tied together. Each chute was in a row by itself. We had three of them. This was next to the north side of the stockyard headed north, and the yards were used for holding pens. We also had a chute for calf roping and doggin' steers. We built an area with netting and a pole on top, and a little corral and unsaddling chute at the end of the arena.

We brought down a few horses from the Hills, including a gray mare, Queen, who was always pretty bronky. On our ranch she was used at haying time on the mowing machine. We also got a string from the Browning country. Some had been bucked, but most were green horses that hadn't been tried. Most of these horses belonged to Wild Horse Johnson, an uncle of Deloris House. Spud Johnson, Deloris' father, rode saddle bronc at the rodeo.

Mabel Mae, my sister, and Elsie Wehr, my future wife, sold and collected tickets at the gate leading into a pasture before getting to the arena. Our bronc and bareback riding judges were Howard Sailor of Gildford and Johnny Brinkman from the Marias River. I don't remember now who our flagman was on the timed events, or who was the stopwatch timer.

We drew a lot of good riders from Canada, the Bear Paws, Flatness Brothers from Big Sandy, Joe Goose from the Rocky Boy Reservation, and also Roy Houle from Rocky Boy. Roy was one tough Indian to buck off and he rode saddle broncs until he was getting pretty gray around the bottom of his hat.

The first day I drew Queen. Hell, I knew I couldn't ride one side of her let alone both sides. Ernest Price had tried her at home before taking her to the rodeo, and he just didn't get her rode. After my Dad found out I'd pulled Queen's number out of the hat, he said to me (I can remember it as if it were yesterday) "Now don't go out there and get too wild on that old mare, 'cause she is going to buck you off whatever you do". And she wasted me real hard. She got rid of all her passengers the next three days right in front of the chutes, a swapping ends and sunfishing. She was easy work for the pickup men and there were four cowboys who didn't have far to walk from where they started. Anyhow, the next three days some real good riders drew her and she bucked them all off. A horse like her in today's times would have been worth a lot of cartwheels on the rodeo circuit. But the next week she had her head stickin' through a collar and was helping mow hay on the Demarest Ranch.

I had good luck on my next three draws, except the horse on the last day; they didn't get her picked up. She had quit bucking and was just running around the arena. She slowed up so I kicked loose from her and jumped off and sprained my ankle real bad. Oh yes, and that evening I was to be Best Man at the wedding of Harry Gardener and Velma Wehr. It being a public wedding, it was held on the porch of the old Dance Hall in Chester. Being on one leg, I just had to bow out. Walter Wehr, Elsie's brother, took up the slack for me. You know I should have stayed on that last draw till she starved to death and not jumped off, 'cause at times my ankle tells me so yet today. A doctor in Chester taped it up for me, but the next morning they took me to Conrad to Dr. Powers as I was in terrible pain. He looked at the wrapping job and he sure was mad, and says, "You sure did a hell of a poor job on wrapping this up". I let him know real quick that it wasn't my work. Anyhow, after he got done I could have been Best Man for several weddings, but of course I still had the "sympathy limp".

I don't remember what the tickets cost, but we did have a good crowd all four days. But it was a lot of hard work getting things done. It was hot and dry, but we had fun, you bet! Elsie said, "Mabel and I had pockets full of money and we felt rich! Then we subtracted out the expenses and it was all gone, but we sure had fun!"

Claudie, July 3rd, 1934

Carl Iverson on Snip

Chester Rodeo, Claudie on July 2, 1934

Chest stockyards, rodeo chutes and arena

Spud Johnson, rider, and Johnny Brinkman, judge, on Speed

Anthony Fey, our neighbor, bought eight or ten horses from Frank Eggers, several miles south of Chester. This ranch is now owned by the Harmons. Another fellow went with me to trail the horses up to Feys. I knew Mr. Eggers well and when we arrived he told us to put our saddle horses in the barn, and come to the house. He said that his wife was in town but she would soon be home to make supper. A skunk had sure enough left his calling card. Boy it was terrible, but he never mentioned the smell. When his wife arrived she noticed it right off and said, "Frank must have killed it in the house". Frank says, "I don't smell anything." Long before this I'd sure enough decided that one of Frank's set of five senses wasn't working.

The next morning, leading one horse with the others following, we started for The Sweet Grass Hills. We were in a terrible dust storm for several miles just around Chester. Up by Haystack Butte we were in quite a flurry of a snowstorm, but it didn't last very long. We were about a mile south of Whitlash and it was getting pretty dark by then, a cloudburst poured down on us accompanied by an electrical storm. There wasn't any thunder or any lightening flashes; but there must have been a lot of electricity in the air. From my nickel saddlehorn, which is close to my belly button, and up my horse's mane and to the points of his ears was just like little balls of fire. All the other horses were "lit up", too. It is hard to explain. I just threw the lead rope at the horse I was leading and let him go. The horses were all ahead of us by then. The City of Whitlash had one floodlight at Burnham Murray's "Jot 'em Down Grocery Store". We just let the horses go and beat it for home just a half mile north.

All in one day, we were in a duststorm, snow flurry, cloud burst and electric storm. I guess that is Montana weather, a little uncertain. Most cowboys leave town all lit up, but these two cowboys came through town already lit up. Our ranch is just out of the City Limits of Whitlash, just far enough that we don't pay any city taxes and I'm pretty darn sure the town isn't going to spread toward us!

Sanford Price had a horse ranch south of Whitlash. Every spring around March I would help him trail two carloads of horses to Chester.

They were put on the rail and sent to Pennsylvania, so you can see, I've pulled my saddle off quite a few mounts that left the Big Sky Country headed toward The Statue of Liberty. The workhorses were sold, and the saddle horses went to a riding school where they were rented out for pleasure riding.

ONE STUBBORN HORSE

In the fall of 1936, Parsells hired Matt Furnell and I to trail about 30 horses to Devon, to be delivered to a horse buyer. I went up and stayed with Furnells. The next morning we went up the creek west of their ranch and went around high on Middle Butte down to the Cameron Place on the west side of the Butte. The horses were corralled there. We led one horse and the rest followed with one cowboy behind them, so it was a pretty easy trip to Devon. Just as we shut the gate on the horses a fast mail train came zippin' by and bellowing steam and they hit the fence, but it held. We got some groceries under our belts, but there was no place to bed down and saw a few cords of wood, so we headed for home. We had already rode forty or forty-five miles. We stopped on Willow Creek and stayed with Clarence Adams, a brother-in-law of Harold Henry's.

When we left next morning I was leading a horse that he had started to break. All Clarence said was that he was too much horse for him to handle. He was really a big horse. I rode him for over a week before I found out his bad points. I was helping Bob Thompson take beef to Chester and we were down past Wickum's by Haystack Butte and he really came unglued. Yep, and I took up a "homestead", and that road wasn't exactly like a feather bed, but "I never followed the rules, so I never filed" (on my homestead)! When that was over that horse refused to move an inch. We were square in the middle of the road. I remember a car came along and he had to go around us. I was some embarrassed. Anyhow, I rode him the entire winter on Mondays with the First Class mail to Gold Butte. The other days I used a team and sled.

CANOE TO THE RODEO?!

I remember my first time being at Writing on the Stone on Milk River in Canada. Writing on the Stone is called that because there are many pictures carved by long ago Indians into the sandstone formations near the river. We went to a picnic and rodeo. The picnic was on the north side of the river and the rodeo grounds were on the south side. Everybody had to ride in canoes to get across the river to the rodeo. If I had known that, I don't think I would have entered the saddle bronc riding, or maybe I would have come in from the south side, 'cause water and canoes sure made me about as nervous as a prostitute in church. And I swim like a rock. Getting belted down in the cottonwood tree cinched on my bronc felt a lot better than getting into that canoe! I met Art Lund for the first time that day. Yeah, and after the rodeo was over, I still had to take that canoe ride back over to the picnic side of the river!

THE MEAN STUD HORSE

The spring of 1937 we sold a team of workhorses to a fellow on Willow Creek, north of Devon. Smokey and Bonnet were the horses. We rode Smokey a "strawberry roan" for several years, then broke him to work after he came a little "stove-up". I rode Smokey on this trip to save a ride home after delivering the team. I cut across country from Grassy Butte area and was taking it pretty slow and easy cause my mount had seen better days.

I was going through a farmer's field and there was a little bunch of horses, and along comes a big stud and he stopped me. Well Bonnet was in heat and she got covered real quick-like. Soon we went and the stud didn't follow, but pretty soon he came and circled us and wouldn't let me go, and of course the mare was fighting and kicking him. This went on for a couple of miles. It really got pretty serious, as the stud tried to fight my transportation and stop me. He was trying to bite and kick my horse and was just missing me with his hind feet and teeth. I could see in the distance there was a wire fence that would get me out of this field, and Smokey was really about played out and here the stud came again. Long before I got to the fence I turned the mare loose, but she was scared and kept with me. By this time my mount was "all in", and I was getting pretty scared too, 'cause he just missed kicking me several times. Luck was with me because there was a gate. I always carried wire pliers and could have cut the fence if I'd had to. That was a time that had I been a pistol packin' cowboy, I know for sure there would have been a dead stud horse. The horses and myself rested for a good while, and we only had a couple of miles to go to this farm.

Anyhow eleven months later Bonnet should have foaled a bay colt; cause she was bay and so was that damn stud.

Dad was waiting at this farmer's place to get me. We went on into Shelby to see Elsie. She was staying with Harry and Pearl Wehr for a spell before going to Great Falls for the birth of our first child, Auverne, who arrived on May 14th, 1937.

CBC HORSES

The story of the Chappel Brothers Cannery of Rock Island, Illinois is a book of it's own. They branded all their horses with the CBC on the thigh. They had thousands of horses in Eastern Montana and some in the Browning, Montana area. The horses were slaughtered for their Fox Farm and also their big trade overseas. Many stories are in print about the CBC outfit. They had a big CBC roundup all over Eastern Montana.

Matt Morgan, a U. S. Border Patrol officer, picked up two broke CBC saddle horses that had gotten into Canada. He was a friend of the family, having been raised in our Sweet Grass Hills. He and my mother had been in grade school together. Matt left the two horses with us the summer of 1931 and told me to ride them if I wanted to.

One was an Iron Gray, the other a Strawberry Roan. One day the Roan caught me "Smokin' The Same Cigarette" with another cowboy and he just flat bucked me off and bruised the inside calf of my leg. I kept it bandaged for some time when I rode, as it hurt pretty bad next to the saddle fender. I rode with just one stirrup most of the time that fall. It had healed pretty good, then one day it was pretty bad and some small splinter of bones came out of the bruised spot. After that it healed up fast and I was using my right stirrup again.

Matt stopped and got the horses late that fall. Despite having been bucked off and hurting my leg, it made me feel kind of "proud" to be riding a CBC saddle horse.

MY ARABIAN

In 1944 I bought a registered Arabian stallion by the name of Altair. He was gray and weighed less than a thousand pounds, and was "Desert Bred" all through his pedigree. At that time I'm sure that he was the only registered stallion in the Sweet Grass Hills area. I kept a record book of the mares he bred and who their owners were. There were forty-six owners of mares, and quite a few of the names were our Canadian neighbors across the border.

Altair sired a lot of real good cow horses. They were smart and had a lot of cow "savie". I only had him for three years and sold him through a horse auction in Great Falls.

NEW SADDLES

To me the sweetest smell was a brand new saddle in a burlap gunnysack in your local post office. When you stepped inside to get your mail, you knew then that your new saddle had arrived, or maybe one for a neighbor. I have had at different times, three new saddles shipped through our local post office. The first one was when I was a teenager. I saved up my bum-lamb money and bought my first saddle, a Cogshall made in Miles City, Montana. The stamping was big stars about four inches from point to point on every part of the saddle. It's number was a "52". All Cogshall saddles had a different number on the front jockey and seat. I bought two other saddles from Hamleys at Pendelton, Oregon.

Altair, my Arabian stallion

Several of my Canadian cowboy neighbors

have had their new saddles shipped to me. I would ride them for awhile, and then they were considered "second hand", making the duty a lot cheaper than it would have been for a new saddle never used. I remember on one of them I left spur tracks across the seat, but it was not intentionally done to make it look used. The 'ole pony just zigged and I zagged!

Dad tells about camping out with a beef herd down by the Amez-Droz Spring, now located on the property of Gary Aklestad, close to Strawberry Creek. When they camped for the night, Floyd Umphrey laid his brand new saddle next to his camp bed, and tied his saddle horse to it with a long rope. Sometime during the night his horse got "spooked" and down through the beef herd he went. Floyd's new saddle got pretty skinned up. Probably didn't look so new or have that sweet smell any more either!

IV / Elsie & Our Family

THE SCHOOL TEACHER FROM NORTH DAKOTA

In 1930 there came a girl from North Dakota named Elsie R. Wehr to teach school at the Hoffner Ranch on Sage Creek on the east side of East Butte. She was traveling by train and as she got closer to Chester she could see the Sweet Grass Hills out of the train windows, and she was hoping that maybe some day she could go there. Hoffners met her in Chester. By that time it was dark and they took her to the Hoffner Ranch. When she awoke in the morning she was surprised to find herself in those mountains she had seen from the train! So I guess she fell in love with the Sweet Grass Hills real quick like. I'm sure if I had lived down on the Flats out of these Hills; my chances on getting my loop on her would have been tough.

The first time Elsie and I met was Decoration Day, 1930 at the Schafer Ranch. I had ridden over to get a bronc to break for Bill. At dinner they set us next to each other, and really did tease and give us a good send off. I rode home to the ranch from Schafers and here were the Hoffners visiting the folks, with Elsie along. I took that bronc and bucked him out right there in front of the house! No one but me knew how close I was to meeting my shadow. The next Sunday was a big picnic and exploration through all the odd formations of sand rock at Bear's Den. I was late for the picnic and everyone had started out climbing and lookin' around. And of course right away, I spotted Elsie sitting on a big rock and hollering for her brother, "Walter, come, help me down from this rock!" And you know what, that cute little school-mom was lookin' straight at me. Well, what else could I do, but grin and be polite and get her safely down. Then I cautioned her to be real careful, because there are a lot of rattlesnakes around. Well, that sure took care of my situation; she hung tight to this cowboy the rest of the day. (Really, I fibbed a little, as we never had a rattlesnake in the Sweet Grass Hills until 1940.) She was sure a good lookin' trim little filly, so I just dabbed a loop on her and she never took up the slack—came a trottin' right up to me. I guess that is what you call 'love at first sight'. Anyhow, we were married four years later.

We were blessed with 62 married years together. We were a good team, "never a slack tug". This cowboy was fortunate that his loop never missed. I bought her a cream-colored palomino saddle horse named Suntan. Elsie rode a lot of miles on him helping with ranch work. I bought her a pair of spurs, 'cause once in awhile Suntan needed "touchin'up in a tight spot". Elsie was a good cowgirl and Suntan a good cowhorse.

Elsie on the Rock at Bear's Den

One year we wintered and lambed a band of ewes for a Great Falls man. The next spring our sheepherder and Artie Morrison with a team and the sheepherder's wagon, trailed them to Chester. It was in June. It had rained for several days, and the roads were just impassable for our car to make the trip to Chester. Dad or I had to be there to bill out the sheep, help load and do the necessary work to get them on the train. So Speed and I made the trip in ankle deep gumbo in four hours. It is a 40-mile trip to Chester, and I wouldn't have done that to any other horse, except him. Neither one of us was too much in love with the other.

Just north of the yards I stopped to talk to Bunk Bartlett who was herding some canner horses. I remember my boots and spurs were so heavy with gumbo it was a challenge getting off to stretch my legs.

This was in 1932 and Elsie was in Rudyard visiting her relatives, the Matt Wehr family. She was leaving on the train that night for Dillon, Montana where she was going to the college to update her teacher's certificate. Bill Schafer was in town attending a commissioners meeting, so I borrowed his Dodge car (and could you guess) I went to Rudyard, and put Elsie safely on the train. Yep, if Speed and I'd taken all day to ride to Chester, I just might have missed the Choo-Choo in Rudyard. Yes, its true that I had to take care of the business of the sheep and putting them on the train in Chester, but my mind was more on getting to Rudyard in time to see Elsie off on the passenger train. Oh it's great to be in love! It took me a lot longer to come home. I just about had a notion to ride south to Dillon and find a job down there close to town a punchin' cows.

Elsie and Claudie

OUR WEDDING DAY

The morning of December 23, 1934, Elsie and I, my sisters Mabel Mae and Helen, and Carl Iverson left home on a cold, cold morning in Carl's car headed for Havre, Montana. We all got spruced up in our wedding "duds" at the Havre Hotel. Speaking of clothes, we had shopped in Great Falls for Elsie's wedding dress. The sales slip totaled $10.50, which included a beautiful velvet dress, a slip and pajamas. I still have that original sales slip.

That evening we drove to the minister's home to be "hitched up". The minister got us all in position standing by a beautiful Christmas tree. I was in my new suit and tie, and Elsie's wedding gown was a beautiful brown velvet with a matching hat. About the time the ceremony was getting serious, I noticed that Carl was still puffing on a big cigar. I nudged him in the ribs with my elbow. I wished then I had left him puffing away, because he put his hand and cigar in his coat pocket. I thought his coat might start smokin', but I guess he somehow got the cigar pinched out. A good thing the ring was in the other coat pocket or it might have gotten a little smokey.

After all the hand shakes and kisses had made the rounds, I un-dallied my rope that I had dabbed on Elsie four years before, and took a chance that she wasn't going to get away. We headed back to the hotel to change clothes as Elsie and I didn't have too much time to waste before catching the train for Minot, North Dakota. I remember Howard Sailor was in the hotel lobby and was giving us all some "fun talk". We got re-packed and hurriedly gets into Carl's car to head for the depot, and it wouldn't start. It had froze up and busted the radiator. In those times Anti-Freeze wasn't heard of and you either kept the motor running or drained the water so it didn't "bust". So we all grabbed a suitcase and headed for the depot, which was several blocks away. We barely made the whistle for departure, but we were on our way for our honeymoon!

As for Carl, the next morning he found out his car was beyond repair because of the busted block. He junked it and bought another second hander to drive home.

Our wedding day,
December 23, 1934

50 years later, 1984

LAST DAY OF SCHOOL PICNIC

I believe it was 1935 and Elsie was teaching at the Schaffer School. Her only two pupils were Lois Schafer and Claude Lane. Elsie wanted to go on a picnic up in East Butte at the head of Big Sage Creek and asked me if I would go to show them the way and help out. You bet, 'cause I loved my wife and liked kids, but as for picnics, it was hard to get me out the back door on the lawn for one. Guess maybe I was kind of a "picnic pooper" on that subject. While fencing, haying and trailing cattle, I ate so many lunches here and there under the Big Sky of Montana on our ranch that I had plenty of "picnics" on a frequent basis. I had to take lunches as our holdings were (still are) pretty scattered and far apart, and often I was not close enough to the main ranch to come home for the noon meal.

It was quite an undertaking to carry picnic goodies on the back of your saddle, and it was quite a few miles to ride. But we did it, and ate at the head of Big Sage Creek. Then we rode over the divide to Devil's Chimney and looked that over. You had to crawl through a hole next to the ground. Inside was a pretty good-sized room in which you could stand up. Overhead was a big hole and through it you could see daylight and Big Sky Montana. This was the "Last Day of School Picnic". The round trip was about 25 or more miles. The picnic goodies carried back easier than they did going to the head of Big Sage Creek.

I always teased Elsie about her taking a cup along to a picnic to dip

Claudie, Lois Schafer and Claude Lane

water from a stream or spring, when she could have just "bellied up" to that water. But at this picnic they got a snapshot picture of C. S. D. sitting by the water, drinking out of a cup. Lois is really laughing, and me a little embarrassed about getting caught. Anyhow, I still belly up to those creeks and springs for a sip of ice cold water.

And by golly, you know, I had as much fun as the two kids and their teacher. Another thing that helped (make my day) was the horse I was riding, Skitter, belonged to Dave Thomas and needed those miles and climbing over the divide to Devil's Chimney.

MR. FEY'S GOLD BUTTE BIRTHDAY PARTY

Saint Patrick's Day, the 17th of March was a big day and night in Gold Butte for Tony Fey Sr.'s annual birthday party. A red letter day on a green St. Patrick's Day! Elsie and I saddled up one morning and rode to one of those birthday parties. It must have been about 1936.

As we were nearing Gold Butte we met George Jackson, Mr. Fey's sheepherder. He asked if we had seen some stray sheep, and while talking he was looking straight up in the sky. We told him we hadn't. As we rode on, we sure knew he wouldn't locate them where he was lookin'. We later found out that's the way his eyes had always been. There is a name for this condition. George Jackson in later years had a little shack where he lived, and by that time there was very little left, people or buildings in Gold Butte. However, it was still a post office and the mailman, Bob Pettapiece, left the Gold Butte people's mail at George's shack for them to pick up. Upon arrival one mail-day the mailman found the shack burned to the ground, and Mr. Jackson perished in the fire. From then on the Gold Butte area ranchers got their mail in Shelby until a mail route came up from Galata.

Later during the years of 1936 and 1937 some people who had faith that there was more gold to be had in Gold Butte came in with big machinery and piled the dirt pretty high. They dug deeper than the pick and shovel miners, who had washed out a lot of gold dust in the early 1900's. I have been told these people with their machinery did pretty well.

Gold Butte had a general store owned by the Weideman family, that carried about anything you wanted to buy. In 1940 they had a big all day sale. I remember us going and Doug was just a baby. Elsie and her sister Velma Gardener bought Auverne and Shirley old-fashioned little high-topped black shoes that buttoned up the side. Velma laughs when she tells about the new student minister who came to the Demarest Ranch one day that summer, and Auverne and Shirley in their little short dresses were running around in the high-topped button shoes. That student minister must have thought he was way back in the boondocks if the fashions were that out of date here in the Sweet Grass Hills!

But that was a few years after this party we attended. The party was at the Fey home, which was the Gold Butte Hotel. We wined, dined and danced till the wee hours of the next morning. As I remember a piano player and a fiddler furnished the toe tappin' music. The nine miles going home was a little downhill and the sun was just coming up over East Butte. My, my, a pretty late night out for being married for a couple whole years! But we really did have a big fun time.

MAIL DELIVERY, TRAPPIN' BEAVERS AND A NEW DAUGHTER

The winter of 1936 and 1937 was a real tough one. That was the winter I drove the stage mail to gold Butte. On Mondays I rode the prior mentioned stubborn horse with just the first class mail; the other five days I used a team and light sleigh. Our mailman had a Model A Ford equipped with caterpillar track for traction, and skis that were placed just a little higher than the front wheel on the car, so when the snow was too tough for the front wheels, the skis did the job. He hired me to take the mail from Whitlash Post Office to the Gold Butte Post Office. His outfit couldn't handle the steep climb into Gold Butte. He usually never got to Whitlash before late in the afternoon. My trip was two hours up, one hour in Gold Butte and two hours back. A good share of the time I was coming back after dark. This lasted around five months. I sure walked plenty of miles that winter in the sleigh track; anyhow I still got all my toes and my nose. That was a busy winter as we had cattle to feed, and I trapped quite a few beavers that winter. I really looked forward to Mondays when I rode horseback with just the first class mail. On the trip home when I had the sleigh it was loaded with groceries for neighbors. The Whitlash store had run out. On the way to Gold Butte I delivered to three mailboxes. Each outfit had their own mail sack, which I put in the boxes at the Johnson Sheep Camp, Glen Aikens and Parsells. Them thar days on this far away route, mail was delivered six days a week. Now it is three times a week. Herman Swan had the Chester to Gold Butte mail contract at that time. He was paying me $6.50 a trip. That was big bucks at that time. Hired ranch hands then were getting around two dollars a day and in the wintertime "Rode the Grub Line" or worked for their board till spring work opened up.

In the spring of that year, 1937, our daughter, Auverne was born in Great Falls. Elsie wrote to me saying her room and board for 12 days in the hospital was $60 which included the baby delivery, medicines, and the works. She was really upset about such a terrific bill. As I said, I had been doing a lot of beaver trapping that winter, and they were bringing a good price. So in other words I guess we paid for our new daughter with a sixty dollar beautiful beaver hide. I remember about that time all the Royal Queens and Society Gals were wearing beaver stoles and coats instead of mink, so that helped beaver prices.

Claudie, Johnny Morrison, and Auverne on DuPont

For years I trapped alot of beavers. It was time consuming work, as I had to tramp down the creek about midnight and also early in the morning checking traps. Otherwise the beaver breaks his leg, and also chews it off and gets away.

A beaver I trapped in 1962

ELSIE, MY COWGIRL

During the years from 1930's to the 1950's there was a shortage of teachers, and Elsie was always in demand. Even after her teaching certificate had expired, she was called on to do a lot of substitute teaching until they could get a new teacher. One of the schools was between Whitlash and Chester at a homestead cabin north of the Hawk's residence where she taught for five months at $60 a month. Her pupil was Leo Hawks. Auverne was a toddler and Sena Hawks took care of her. Another of the schools was down by the Blair Ranch on the Lothair Road. She also taught at the Schafer school several different times. When Auverne started school, Elsie was her teacher at the Bear Gulch

School for her first and then second grades. These were one-room country schools. In the 1950's she had B. J. Oswood one winter as a student, holding his classes in the bunkhouse (which was fixed up real nice) on the Oswood Ranch. Yep, I would come home quite a few times and my Working Partner, Chief Cook & Bottle Washer, and Cowgirl, was over the hill someplace doing what she loved to do - teaching school.

In 1947 when Auverne was old enough for 4-H, Elsie started the Sweet Grass Hills 4-H Club. Irvin Brown was the other leader in the beginning. She was a leader for years and was also on the State 4-H Council. She started the ball rollin' getting community interest and donations for the new calf barn and sales ring building at the Shelby Fairgrounds. This building is still being used today.

She gave many piano lessons to children of the community. Elsie was also active in the Whitlash Ladies Home Demonstration club. We both belonged to the Whitlash Presbyterian Church and Elsie was Secretary-Treasurer at the time we started a fund for a new church building. On January 28th, 1945 we took up a collection of $7.80 to start the building fund. This helped the building fund grow. By February 1952 the church was built and paid for from donations and with some help from the Board of National Missions and (all told) the church wound up with a balance of $254.20. The church in Whitlash had it's own registered cattle brand. In the spring the ranchers would brand a calf for the church and run the "Church Calf" along with their own herd. In the fall the church received calf checks when those calves were sold. This goes to show what a "Hill Billy" community back in the boondocks can do when they all work together, "Never a Slack Tug" and all pulling in the same direction.

Now and for many, many years to support the church, there is an annual fall dinner followed by an auction of handmade items. Folks from other neighboring communities come. Usually every year I buy an afghan or two. This year, 2000, there was a beautiful quilt with horses and brands on it. I quit bidding at $750. It sold for $1000 and had been made by Jeanette Brown and was purchased by Darold Tomscheck.

Elsie also loved studying and collecting data about Montana History. One of her projects was collecting local history of the Sweet Grass Hills and all of the people who have lived in this community. She interviewed many of the "old-timers" during the 1950's and 1960's. She also gathered data on the Wehr Family and was very interested in her roots, keeping contact with many aunts, uncles and cousins. From all of the Wehr information in her files, "The Wehr Family History and Genealogy 1816-1997" has been compiled and printed by our daughter. She was an avid reader, especially of historical books.

Both of us were members the Eastern Star Chapter of Galata. I also belonged to the Masons, and took my turn in the East. That was quite an experience.

Elsie died in early October 1996. Her funeral was at the Whitlash Presbyterian Church on October 8th, a beautiful, calm, warm and sunny fall day. At the graveside after Elsie's funeral, I went and shook hands with her pallbearers. They were Ken Morrison, B. J. Oswood, Ron Wehr, Albert Fey, Earl Thompson and Bob Dafoe. I shed a happy tear to notice that they were all cowboy booted for my cowgirl who was my working partner for sixty-two happy years.

Elsie my Cowgirl, and me

ELSIE AND SUNTAN HELP OUT

During World War II, the country was pretty short of ranch help, so Elsie and Suntan, her palomino horse, cowboyed with me. That fall Anthony Fey and we sold our beef to a buyer at home and weighed them at the Government Scales at Whitlash. The buyer wanted the cattle in the stockyards in Chester in two days. We put them in a field at Hawks the first night, and stayed with the Hawks family. The next day we had to cross over a wooden bridge which was pretty "spooky" for cattle. It being a cattle underpass, we could not go around it, the fence on both sides was tied to the bridge. So we just had to be patient till they decided to go across. Down over the hill a lady lived and whenever she saw cattle go by she was always afraid they might water at her reservoir, and that was a "no-no". Anyhow, she came storming up the hill a foot, petticoats a-flying. I motioned her to go back, but on she came, scared the cattle and they made a pretty good hole in the fence. I told her I would patch up the fence. She says "I know you will, 'cause I'm going to stay here to see that you do". I patched the fence and Elsie went on with the cattle by herself. So we lost some valuable time.

Just the two of us could have had trouble corralling them if it got dark or a fast mail train whistlin' and kickin' out steam came whistling through town. We were lucky though; no train came through and we still could see. Elsie had the reputation of being one of the best school-teachers Liberty County ever had. But put her on a top cow horse, she was one hell of a hand. She proved it on that drive because the Fey steers were three-year olds and very nervous to handle.

Elsie and Suntan

They would just about spook at their own shadow. I rode a Fey horse named Roller, one of the best horses that I ever saddled and threw a leg over. He was solid bay in color. The Fey steers carried the  on the shoulder.

REGISTERED HEREFORDS

My start in the Registered Herefords was with a registered Hereford cow that was in a bunch of commercial cows we bought in the fall of 1940 from Brinkmans at Marias River. Miss Stanway 30th calved July 3rd, 1929; Breeder, Elk Run Ranch, Highwood, Montana; Sire, Imperial Stanway; Dam, Quinto 4th. In 1944 I added a few cows from Archie Parks. Also in 1945, I bought cows from Ox Bow Ranch at Wolf Creek, and also some heifers from Savik & Larson. For a short time I used a Stephenson & Hoover bull; Blanchard Domino; Sire, Prince Domino 9th, Dam, Martha Mischief. In 1949 we bought the Grand Champion bull in a breeders' consignment sale held in Great Falls; Teton B. Mischief, Sire: Battle Royal 6th bred by Savik & Larson of Bynum. He was a great bull, crossed exceptionally well on any bloodline cows. His calves were our foundation cowherd. We sold bulls at the Marias Hereford Breeders' sale and also consigned bulls in Chinook and Malta, and Private Treaty at home.

We also used a Real Silver Domino, bred by Hunter Brothers in MacLeod, Alberta, Canada; also an Errol Harland sire bred by Higgins Brothers, Ringling Montana. At the Winter Fair at Bozeman I purchased the Reserve Champion bull, V.H. Proud Prince 6th sired by Proud Prince, bred by Yellowstone Hereford Ranch at Miles City. He turned out to be an exceptional cross on Teton B. Mischief heifers. Other bulls purchased was a herd bull from Eaton Becker at Willsall, Montana, and a herd bull at the Calgary Bull Sale. There were 700 Hereford bulls at that sale and it took three days to sell them. I stayed all through the sale and Elsie went with me. She visited some relatives while there. We just had a half-ton pickup with a homemade stock-rack. We barely got the bull poured in and the endgate closed. He fit pretty tight. Some ranchers gave me a bad time and kidded me about raising bulls and still going all over the country spaying heifers.

Claudie at Bullsale

Showing bull at Cut Bank

Registered Bull

I helped form The Marias Hereford Breeders Association at Shelby and took my turn as President. We were the first in the state to have a Bull Indexing Program, to check the rate of gain per animal. They were weighed in at the start and after. At present all the breeders index most all registered bulls.

WORKING AT THE STOCKYARDS IN SOUTH SAINT PAUL

Trainloads of livestock were shipped east every fall on the Great Northern Railway to the Terminal Markets in Chicago, South St. Paul or Sioux City. Soon after that a few Auction Sales Rings were starting to be popular, and order buyers were placing cattle direct to the feeders in "The Corn Patch" from the auction rings and also from the ranches, but the livestock still went by rail to their destination. Around 1970 The Great Northern discontinued hauling livestock. Since then trucks have taken over the transportation to the feedlots in the Midwest.

In 1944 or 1945 I worked for Charlie Govern in the South Saint Paul

Terminal Market. He owned a commission firm and always tried to get country cowboys to work in the fall run of cattle. Charlie called me and Dad knew it would be a great experience for me. The same year Buck Boyce from Havre, and Bob Cardwell from Billings also went. Bob was married to Charlie's daughter Muriel who worked in the office. I worked with the brand inspector, Loren Daley, on Montana cattle. Artie Schanahan was the brand inspector for the North and South Dakota cattle. I worked two different brand books. When these cattle were sold and went to the scalehouse to be weighed, my job was to call the brand on the animal and the owner's name into an opening into the scalehouse to Buck Boyce. Buck was making out the tickets on the brand and ownership. Buck was a son of Steve Boyce who lived in the Bear Paw Mountains south of Havre, Montana. There was a "Tally Sheet" that came from the loading point to destination. It came to the Montana State Brand Inspector. We knew what brand to look for and it sure helped. But the North Dakota and South Dakota cattle didn't come with a "Tally Sheet" to the Brand Inspector. I helped the brand inspectors rope and clip many cattle whose brands couldn't be read when in the Main Alley. I came home and told everyone to be sure and brand them big and plain and use an iron with a little wider face.

The South Saint Paul Stockyards was a "beehive" of people and livestock. Trainloads of cattle came there every day to be consigned to commission firms. There were empty pens left unlocked where the railroad crew unloaded livestock from the cattle cars. Then the pens were locked. Later they were taken to the Main Alley and sorted for sex, size and condition, then brand inspected to determine ownership. When the Brand Inspector came to work the cattle, first the pen number was written down in the Brand Book. The cattle were then counted out of the pen into the alley. They were then let back into their pen one at a time, and the critter's brand was read and marked down in the Brand Book. The hardest part for me to get used to was learning to move around and write in the Brand Book at the same time. While working in the yards that fall I rode a brown gelding that had come from Kill Deer, North Dakota. I used him taking cattle from the pens to the scalehouse, saving me a lot of walking.

Local finished beef cattle and dairy cattle were trucked in to the different firms. Each firm had an office in a large building called "The Exchange Building". The Charlie Govern firm had a different crew located at the unloading chutes for the local fat, finished or dairy cattle. No brand inspections were needed for them. Swift and Armour Packing Plants were just a few blocks away, and they employed lots of people. Down in the yards the owner of the Firm went into a pen of cattle with either a packer buyer or a country feeder buyer and got together on a price. The fat cattle going for slaughter were taken from the yards to the packing plants on an overhead ramp. I took several tours of the packing plants, once with Doug and Tommy Lakey. We watched the complete process from the killing of the animals, skinning, gutting to seeing the carcass hanging on the hook.

While I was there I got a call from the Hamilton Ranch near Simms, Montana, wanting me to spay 1500 head of heifers. When I told Charlie about the call, he kind of rolled his cigar around a few times and says, "Well, you better go. If it didn't have something to do with livestock, I would say 'no way'"! So I caught the train to Montana, got the job done and then went back to the "Main Alley" at the South Saint Paul Terminal Market. I saddled up the brown horse and started lining up cattle for the scales.

I have to tell about Loren Daley's "Sliver Trick" that he pulled on ranchers, cattle buyers and feeders. He had a real thick callous in the palm of his hand caused from roping so many cattle in the stockyards in order to clip their brands. He would take a straw from a broom and run it back up his shirt sleeve so it didn't show, and then he would push the end through about an inch of callous, and let it barely stick out. He would tell someone he got a bad plank sliver, and would they get hold of it to pull it out, but he held his free hand over the other and pushed down on the "sliver". He would hold tight and after a few pulls and jerks he would release the pressure and out would come about a foot of broom straw! One day up at the eating place in Ollers Bar and Café, he approached a couple of ladies in a booth and told them he had a big sliver and could they get it out for him. One lady took a look and turned pale and said that she just couldn't, but the other one says, "I can get it" and digs in her purse and comes out with some tweezers. She also put some disinfectant around the sliver and Daley let her work pretty hard before he let her have it, and then out comes the foot long "sliver"! The first lady about fainted and the other brave one turned a little pale. I have seen Corn Patch Buyers, Ranchers and anyone who pulled out Daley's foot long sliver have a pretty startled look, and Loren's face looking like he sure was hurting. Then he would laugh and laugh, no matter how many times he pulled this joke on some unsuspecting person.

When I came home on the train to Shelby from South Saint Paul I caught a ride out to the ranch with Dot and Red Morris. They were hauling a bull in their pickup. There had been a blizzard and the Strawberry Road had some big drifts. One was on the bend south of Ron and Toni Wehr's turnoff (the former Furnell Ranch). At that time this road had a real bad dip and it was full of snow. Red took a run at it and really buried the outfit. While Red had the hood up and was digging the snow away from the drain plugs so he could drain the water out of the radiator, I opened my suitcase so Dot could find herself some of my clothes to put on. We were going to have to walk across country to their ranch about two miles away. She didn't have enough clothes on her to flag a handcar. While this was going on I was getting the bull unloaded so he could travel around and keep himself warm. Dot put on a pair of my extra Levi's, another coat, and my overshoes. When we started out, Red was kind of dizzy and couldn't walk by himself, so we held on to him for two miles while we trudged through snowdrifts and it was getting dark. Fortunately it wasn't storming. Red had evidently breathed too much exhaust fumes while lying over the motor under the open hood of the pickup. We were all in and just barely made it to their ranch. By then it was dark. As soon as we got into the warm house and relaxed, Red came right out of this spell. He had been dressed warm. I had about as much snow as feet in my boots. To keep warm I had put a scarf over my head and then slapped my Stetson over the top. I stayed at Dot and Red's all night. The next morning Dad came up leading a horse for me to ride home.

I was glad to get back to the Sweet Grass Hills and my family. Elsie said that Doug had really missed me, and cried a lot wanting me to come home. It was a great experience and I met lots of new friends working at the South Saint Paul Stockyards. I also got to know many of the cattle feeders and ranchers from North Dakota, South Dakota, and Montana.

PET DEER AND ANTELOPE

When Auverne and Doug were small and growing up, we had at different times two mule deer fawns, both females. I brought the first one home from up in the Hick's place. We named her Faylene. She drank out of a bottle, and then the next spring she started getting a little choosy on her diet, and visited around the community eating in the vegetable gardens and flower gardens. She played with the kids, following them down the hill on their sleds, and in the summer (they went to summer school in Whitlash) followed them to school and lay under the schoolhouse windows. One day they heard a little "tap, tap, tap" of her hooves, and here she was, walked into the schoolroom through the open door. When strangers came to the ranch, she got excited and ran and ran around in circles around the house. She began to jump up with her front feet on the kids' backs. Finally she became such a problem, we sent her to the Game Reserve east of Malta, Montana. The caretaker came up to get her. Harry Gardener and his family were here, and he built a crate to put her in. My Mother had a little dog called Simpie, part bull and terrier. He and Faylene were good friends. When the crate was ready for her journey, Simpy tried his best to tear it apart and get her out, biting and pulling on the wooden slats. When the car left with her lookin' out of her cage from the open trunk, the kids were all cryin' and Simpy followed the outfit up the road quite a distance before giving up and coming back home. It was a sad day for everyone. Faylene became a pet at the Reserve for the caretaker and his family.

The other fawn was with us one summer and into winter and left us. But she came back the next spring and was so happy to play with our dogs and pester us for milk and goodies. She stayed quite a long time, than finally, I guess went back to her own kind. We called her Kid.

Both these deer were real friendly, also with our dogs and ran and played with each other. Could hardly keep them out of the house.

Our next pet was an antelope, and he really loved his bottle of milk. When he got a little older, Doug would go out with his bottle and lean up against the house, then would call him to come; otherwise he would about push him over, if not leaning against the house. Later into the summer he got a little horny, wanting to push the kids around, kind of rough-like. I say to Sid House, "Lets try and castrate him". By this time he was getting pretty good sized, so while Doug was holding a big bottle of milk and him a suckin', Sid laid him down, holding him by the legs. We had doubts about what we were about to do, and really didn't think we could get the job done until we were through with what was going to take place. However, he just laid there without a struggle, and he just sucked, sucked, sucked on that nipple, and real quick like I relieved him of his Family Jewels, and throwed a little sulfa powder on the operation. Just barely got through and he was on his feet. I can't yet believe that the way we planned it, that Doug didn't run short of milk first. Guess that bottle was far more precious than the hurt. He was a perfect little gentleman from then on.

Doug, Faylene and Claudie

That fall after Auverne and Doug went to school (in Shelby) and were gone, he got to going up to the gas camp as there were kids up there to give him some attention, also they were feeding him some milk, but he still came home every day for more. That fall some hunter shot him and just left him lay. His name was Rickie.

Another year Doug carried a little baby antelope home on his saddle horse; said he was so afraid of hurting his little fragile legs while carrying him. But he was doing good and really loved his milk from the bottle. However, he developed sores in his mouth and died from this, 'cause he could not take any milk. So we lost him.

One time when fencing up at the Hick's Place, I found a baby mule deer fawn that was pretty hungry, so I told Earl Thompson about it. They went and got it and had it for a couple of years.

AUVERNE AND DOUGLAS

Elsie and I had two kids, Auverne and Douglas. They were both born in Great Falls, Auverne in 1937 and Doug in 1940.

We were driving down the country road when Auverne was a toddler and went by a lake with a duck swimming. "Oh, oh," she yells, "Chicken on the Lollie!" That's what she called water. Well, she grew up and went to high school in Shelby, and went to Northern Montana College in Havre and became a medical secretary. She then moved to Seattle and worked there for several years and married Tim Dieffenbach who was a teacher. Later when they lived in California she went back to school and became a Registered Nurse. Tim taught and coached in a Junior High School. She and Tim raised their family in California, but moved back to Washington State in the late 1970's. Tim retired from 35 years of teaching and coaching junior high and high school a few years ago. Auverne recently retired after twenty-seven years nursing, twenty-five of those years in Critical Care.

Oh, I just thought of a something that shocked my son-in-law, Tim. They hadn't been married very long, and this Montana country life was new to him, being he grew up in the city of Seattle. One day he and I were in Shelby at the sales ring. Another guy and I were out in a pen of my cattle looking them over and Tim was sitting up on the corral fence. As I walked by, one of the critters raised her tail high and I got a shot of fresh manure all down my left shoulder and coat sleeve. Tim said that I just kept walkin', talkin' and lookin' at the cattle, like I never even noticed what happened. He said to me, "Claude, how can you stand that! That cow just crapped all over you!" I just pointed to my left shoulder and said, "Gold dust, Tim, gold dust".

Auverne comes out to the ranch quite often and always to "ramrod" the crew at branding time, cooks the branding dinner, and has a good time with all the "natives" where she grew up. In her spare time, she is typing away on her computer, as she is my publisher for this book. She is still a "Little Bit Montana", 'cause she loves living out in the country and spends a lot of time diggin' in her woodland garden and her flower garden. They live on a lake and now she still sees a lot of "chickens on the lollie"!

Doug, for so long after he learned to talk could not pronounce his F's, they all sounded like a W instead. One cold day he came in from outside and stood in front of the little gas-stove heater and said, "Wire is sure a good ting!" Another time bringing his Mother Goose book to Elsie he said, "Wead me the story of the wox if you can wind it." He talks better now!

Doug was home from high school in Shelby every weekend and he spent many of those weekends up at the Hick's cabin as we were feeding cattle up there in the winter. He started breaking horses to ride at a young age, both for our ranch and for other people. Doug started roping real young and usually ropes at all the brandings.

Doug rode bareback broncs at some of the nearby rodeos. His rodeo pals at that time were Cliff Forseth, Ken Morrison, Clarence Morrison, and Jim Fersch. At the Choteau rodeo a fellow remarked to me that his draw was a little rank, and I said, "Well, he might not get her rode, but she will know she got a good spurring!"

He went to horseshoeing school in California, and also became a pilot. He had a lot of pleasure with his 150 Cessna Airplane. We had a good landing strip and windsock out in one of our fields. Some years later he sold it for twice what it had cost him (the airplane, not the windsock!)

Doug took his turn in the Army and was on the Military Police Force in Fort Carson, close to Colorado Springs, Colorado. While he was there he had his black and white International pickup with him, so he got to see a lot of that area. When his time was up in the service, they

Claudie Elsie, Auverne and Doug—1945

asked him if he would consider going to a Leadership School. His answer was, "All I want to do is go home and lead a horse to water".

He has been on the ranch all his life, raised his three kids on the

Doug

Doug on Prince, roping—1945

ranch, and has now taken over the reins from me. He is a good cook, fixes meat, potatoes and veggies. We usually eat the noon meal together at his house. Sally, my little white poodle keeps me company, never leaving my side, unless I go to town and have to leave her home. Most of the time she comes with me, though. Jim, Doug's big black dog lives at his house and helps take care of everything outside too. Jim is runnin' and checkin' every inch of the ranch several times a day, flushing out deer, porcupines, skunk, birds, or whatever else he finds along the creek in the brush. He is quick to let us know if there is anything unusual around.

Doug, Claudie, Elsie and Auverne—1995

MY GRANDCHILDREN

I have six grandchildren who are all grown up now. Auverne's children, Sandra, Craig and Brian all spent some time here during the summers. Doug's family, Paula, Marty and Crystal of course were all raised here on the ranch. These are some of my memories of their growing up years.

Of the California grandkids, Sandra, who goes by SanDee, spent the most time here. I remember one day when she was about four years old; she walked down to the chicken house to gather eggs. She had to walk through some tall grass, and saw a snake. Back to the house she came, screaming bloody murder. She went upstairs and got on a pair of my boots which came up past her knees, and went tromping back to the chicken house. She was determined to gather those eggs! She spent one school year with us and rode the school bus to Chester High School. She had to get up extra early to bottle feed her little calf before school. She loved horses and to ride, and most of all going to horse shows and competing in all the events. She took her share of ribbons in barrel racing and other events. One year she was the high point winner in the local Saddle Club for her age group. She was thrilled to compete in the Regional Finals in Cut Bank. The day before the competition her horse got into a kicking match with another horse and ended up with an injured leg. So SanDee rode a big appaloosa horse at the Regional Finals. Jim Cochran loaned her the horse. She helped me a lot with ranch work and trailing cattle. She enjoyed working in the garden, and was a great help for Elsie in the kitchen. She learned to do a lot of cooking and baking at an early age, and she is a really good cook. I can't take credit for that, her Grandma taught her. Her husband, Mike Orman is a carpenter. She is now a busy Mom and a Registered Nurse working in a hospital critical care unit. Their daughter is Tiffany Orman.

Craig is a couple of years younger than SanDee. One day he was with me out in the field and he wanted to catch a gopher. I gave him a piece of string and showed him how to make a snare and left him sitting at a gopher hole with the snare. I started off down toward the shed, thinking, "That will keep him busy for a long time." I hadn't gotten far when I heard, "Grandpa! Grandpa, I got one!" Sure enough he did. Another time we were at the Hick's Place and he wanted to fish in the reservoir, so I got him set up with pole and hook. I'll be darned if he didn't catch the biggest fish ever taken out of there! He was eight years old and got a write up with a photo of him and his fish in the Chester paper that summer. Another time he was with Doug and me riding horseback up high in East Butte. He decided he had to have a drink of water and begged us to let him ride a couple of miles to the creek for a drink, and of course we wouldn't let him go. He got pretty mad at us and said, "It's a regular desert up here. Look at those two trees over there, they are even dead!" When fencing with Byron and I in Middle Butte he was always finding a rock that he was sure had gold in it and he would be so excited he would almost take off running down the mountain with his find. He was a patient and persistent kid who believed he could do anything, and by the sounds of it, he continues to have that attitude as a successful entrepreneur and businessman. When Craig was just a little guy, he pounded a nickel in a fence post and held it in with a fencing staple. He said, "Grandpa, now you will never be broke." The nickel is still there. Craig's daughter is Giovanna Dieffenbach.

Brian is the youngest of the three Dieffenbach grandchildren. When he was five years old, he made a business deal with me. He too liked catching gophers, but I think he trapped most of his. I had to pay him five cents for every gopher tail he turned in to me. His gopher catching was a real thrill for him, and Grandpa had to keep a pocket of nickels and pennies real handy. The pennies were for each dandelion he pulled and for each potato bug he caught. During an O'Mox'see at the Brown Ranch, Tim and Auverne drove up to the arena fence having just arrived from California. They were thrilled and half in shock, to see their little blond kid on a big stout horse out there in competition and going full speed. He was a good little rider, and stuck to the saddle like glue. He also took his turn at riding calves during branding. Brian liked to ride horses, and even though Inkspot bucked him off he got right back on. He really liked to go fishing and caught many in the Fish Lake, which he helped stock. He and Craig spent hours out in the rubber raft with the mosquitoes biting, letting loose the little fingerling trout after they got used to the temperature of the lake. Some years later he helped clean out dead fish that had died over the winter; one was the size of six and a half "frog-skins". Since he liked to fish so much, I taught him how I used to fish in the creek without a pole, just catch 'em and throw 'em out on the bank. He is now manager of an Italian restaurant. Brian's daughter is Amanda Dieffenbach.

Paula came to the ranch when her mother and Doug got married. She was five years old. Paula was a horse lover, too. Her folks bought her a small built horse called Trixie. She punched cows on her and also entered her in horse shows. We built her an arena here at the ranch where she could practice riding English and jumping her horse. Later she learned to play polo. I helped her learn to drive the pickup out in the field where we had plenty of room and nobody watching. Oh yes, when she was learning to drive I remember we had a few gates to repair! One winter, my nephew, Claude Ritter was here and trapped a bunch of beaver. Paula would always get up real early and go with him checking traps. She loved it! Two different years as we were coming to the corral with cows and calves to brand, Trixie came over the hill with an empty saddle and Paula was walking. Must have been a buckoff. She still rides and is busy being a Mom to two very active little cowboys, Reid and Rylan.

Marty was pretty young and was with Doug and me. We were driving down the road and he was for miles and miles really a chattering, not giving us any chance to visit. Finally Doug said, "Oh Marty, you talk too much." Marty's reply, "Well, that's just the way God made me", and he kept right on chattering all the way into town for 60 miles. One day he was at our house, and outside it was a downpour of rain, a "Regular Toad Strangler". He wanted to go home and Doug told him not to, that he would get his clothes soaking wet. Pretty soon we looked out the window and there ran Marty without a stitch of clothes on and carrying them in a paper sack, headed up the hill to his house! Marty was the bookworm and musician on the ranch. When he was very small he went to the Library in Shelby and brought a stack of books to the counter to check out. The Librarian said, "You can't check out all of those, you won't even be able to read them." Marty replied that yes, he could read them, and proceeded to read for her. He was allowed to check out all the books he wanted. He and Elsie used to look up lots of things in the encyclopedias, and she also gave him piano lessons. Marty excelled in music, and is an accomplished pianist. He now works for a public broadcasting station.

Crystal Dawn is my youngest grandchild, and has always loved horses, her dogs and the ranch. She won many ribbons showing horses. She also trained her dog, Stretch, and took him to dog shows. In Great Falls he took first in his Class & Breed, and during the same show he took first over all breeds of dogs. One night when she was still under school age, she came over and was going to stay all night at our house. She went to bed in our downstairs spare bedroom. During the middle of the night I checked on her and she was gone. I went to the entry and her snowsuit and overshoes were gone. It was below zero and there was snow on the ground, so I followed her little tracks over to her house. She was already in her bed and sound asleep. She and her husband, Matt Hall, are both attending college. Matt is in diesel mechanics. They come to the ranch most every weekend, so they help us out a lot.

Craig and fish he caught at Hick's Place

Brian and gophers he trapped

SanDee with ribbons she won at horse show

Marty with Penelope, and Crystal with Stretch

(Left to right) Doug, Marty, Paula holding Reid, Crystal, Claudie, Auverne, Tim, Craig holding Giovanna, Brian, SanDee—October 8, 1996

V / Punchin' Cows

SHIPPING CATTLE BACK EAST

When I was around 12 years old, I took my first trip trailing beef cattle the forty miles to Chester. We trailed with Mr. Strode's cattle and the first night put them in a field south of the Hill Post Office. The second night we corralled them in a large shed about eight miles north of Chester. The cowboys were John Lauener, Tom Morgan, Slim Stillman who was a relative of Mr. Strode, my Dad and myself.

At that time the stockyards for loading cattle on the rail was on the south side of the railroad tracks, about where Tiber Tractor Implement is now. The yards would not hold many cattle, so we had to take small bunches into the yards, load them on the stock cars, then get another bunch. Johnny Morgan, our Liberty County Sheriff always kept some saddle horses in town and he always helped the ranchers on shipping days.

I was pretty keyed up as I was getting to go to Chicago with the cattle. At the depot we were given a stock contract as caretakers for two loads of cattle, and between every division the conductor would punch your contract. So when the conductor asked for my contract, he looked at me and said, "You have got to be eighteen years old to ride to Chicago with cattle; just how old are you?" Well, I kind of got 'tall in the saddle' at the moment, looked him in the eye and says, "I am eighteen". He than says, "I know you're not eighteen and I'll have to put you off at Havre." Then is when I didn't feel so tall. Pretty soon he patted me on the back and says, "Oh yes, that's right, your Dad is with you". Anyhow, I'm sure it's the only lie I ever told in my life, and it was a "Big One"!

The trains got pretty long, starting at Browning, picking up cattle and cowboys along the way. At Havre more cars from Great Falls hooked up. The cattle were unloaded and fed in Minot, North Dakota, and at feed yard in Minneapolis, Minnesota. Crews at the feeding points unloaded, watered and fed the cattle and then re-loaded them. The hay charge was deducted from your cattle check when they were sold. Then we went on to Chicago, Illinois to the Terminal Market. After living on a ranch in the country and once in awhile getting to Chester, my neck got pretty stiff looking up at those tall skyscrapers. So it was a great experience for this young cowboy.

On the trip there were a number of livestock men along with their cattle. We were furnished an "old-time" coach for the cattlemen. With all the ranchers and cowboys traveling with these cattle, it got pretty "Western". There were no dull moments with the poker playing mixed with some giggle juice! Everyone had to pack their own food for the trip. There was lots of fried chicken in all those lunches. Lots of roosters had lost their heads back home at the ranch at shipping time! On these trips most ranchers just let their hair down and tried to "buck off their hat". It was Powder River a mile wide and a foot deep, yep, it was.

Our stock contracts were taken up and we were given a regular Pullman ticket for our return home "free" on the Empire Builder passenger train. Our cattle were consigned to Miller, White & Woods, a livestock-selling agency. I was a little bit spooked being in such a foreign environment and pretty well hung on to my Dad's shirttail while taking in the city sights. And my Dad's shirttail was usually hangin' out, no foolin'.

From about 1937 until 1969, over thirty years, I rode the stock train to South Saint Paul almost every year except on a rare occasion when we sold our beef cattle to a buyer at home. We would have to trail them for delivery at Chester. One trip Tommy Lakey and Doug went back to South Saint Paul with the cattle. They got a big laugh sitting on the fence and watching Charlie Govern a horseback trying to make out the brand on a certain cow critter. Finally he just gave up and said, "It looks like a monkey sitting on a pump handle".

Another year Jerry Buckley went back to Saint Paul on the stock train with me. During the years when lots of cattle were being shipped they provided the ranchers with old "retired"coaches for the trip, but in later years when not as many cattle were being shipped by rail, they would put on an extra caboose for the cowboys going with cattle back east. Being the first time Jerry had rode the caboose, he went to the little 2x4 toilet. When he came out his hair was standing straight up, and he says, "That's sure as hell a fool-proof sewer system!" When we happened to have an engineer who was a little rough on the throttle of his engine the whole train would be jerked, especially the caboose. The conductor would say, "We got so and so Hog-Head on tonight". One trip there were two cabooses, the last one for the conductor and the other one for the cowboys. Doug was up in the high seat, where you could look down over the whole top of the train and he saw the derailment before the crew was aware of it. We were east of Minot and two cars went off the track, just ahead of us, but they didn't tip over. The cars' left set of wheels were between the rails and the right set was outside the rails which made for a bumpy ride across the railroad ties. Behind us the grass was on fire in the ditch and the train soon came to a stop. We got our suitcases and got out in a hurry, but the fire got stopped close to us at a railroad crossing. Another time we were heading into a heavy rain. The front and back doors on the caboose had about an inch space at the bottom, and there was a stream of water coming in and going out for quite a few miles.

We had to order our cattle-cars two weeks in advance of the date we loaded. So in those days a decision had to be made way in advance as to what date you wanted your cattle at the Terminal Market. You had to allow the two weeks to have the cars available on the right date, and add eight days which included trailing to town and shipping on the rail. Now days we wonder, "Should I contract these calves today or wait awhile, or go to the auction ring this week or next?" The cattle-cars were 36 feet or 44 feet in length. Sheep cars had double decks, sheep on the main floor and sheep upstairs. The shipper paid for 24000# per car, a rate of 64 cents per 100# at destination. At those figures it cost $153.60

per car of cattle for a 1000 mile ride from Chester to St. Paul, and if the weight was over 24000# you paid extra. When you ordered your cattle-cars to ship, you could order what they called a "trailer car" and only had to pay for 12000# instead of 24000#. That way it was real handy for extra cattle that didn't fit in the loaded cars. Otherwise they would have to go back to the ranch. The shipping charges were taken out of your check when the figures went through the office at the commission firm that you consigned to. Different years the price could be a little more or less.

The cattle were unloaded and fed at Minot, North Dakota; that would be on a good run. During World War II, all war material had priority over other merchandise and we would get sidetracked and would have to feed at Williston, North Dakota and Breckenridge, Minnesota. On those trips you met a lot of ranchers from different areas, and feeders from the Corn Belt. A few blocks away from the South Saint Paul Livestock Yards were Swift Packing Company and Armour Packing Company where the fat cattle went for slaughter.

During the 30's and 40's a lot of sheep were loaded out of Chester, so they started having Wednesday for cattle to load, and Fridays were for sheep, so that the facilities could handle them. Also it made it a little easier on Law & Order, not to have the Cowboys and Sheepherders all in town on the same night lying, and braggin' and whoopin' it up. It did get wild and woolly at some of the local water holes; Western Hotel & Bar, Rockman's Bar & Cafe, and Edgar Balou's Bar & Poolhall, the only place in town with slot machines and pool tables. I know, 'cause I was helping neighbors trail their cattle to town just about every week through the fall run. After loading it was a forty-mile horseback ride home. If it got late, I would stop part way out at some friends (John Dolezal, Clarence Hawks or Hutchins) for the night.

On one of my trips on the stock train, I stopped to feed in Williston, North Dakota. There were several double deck cars of lambs that had been loaded in Blackfoot or Browning. I was watching the crew unloading, and these lambs must have been loaded in a rain or wet snow, and being late in the fall the temperatures were freezing. The lamb's wool had frozen to each other, and coming down the chute some were going one way and the other one frozen to him trying to go the other way. A lot were dead in the railroad car from being tramped on, a sight that wasn't easy to look at. That was a lot of lost fall income and wasted muttonchops.

The stockyard at Chester that is there now was built in 1928 except for the part added on to the west. Leo Milles, a neighbor of ours had teams of horses and fresnoes to move a lot of dirt for six chutes leading to cattle cars. The cattle came off level ground and went up into the cars. The ground also had to be plowed so the fresnoes could pick up the loose dirt. Charlie Swank from Chester had teams moving dirt. It took all summer. Today with cats and dozers it would have been a small job. Years before, of course, the Great Northern Railroad grade for the tracks to be laid on was all done with teams and fresnoes, and wheel scrapers so it took a lot of copper rivets to repair harnesses, and a lot of hay and oats instead of fuel and oil. I'm sure glad I wasn't around during the earlier times of building the grade for the trains. The dirt moving equipment called "Fresnoes" were named that because they were made by *Fresno Agriculture Works* in Fresno, California.

In the fall of 1928 I helped Jimmy Morrison trail some cattle to be shipped out of Chester. We were out of town about two miles and it was getting dark. Jimmy says, "I better ride ahead and just see where those gates are to the new yards". Neither one of us had been there. These cattle were the first of any livestock to be corralled in the new stockyards. It must have been real good 2x6 material 'cause the two large corrals and gates and alleys and also the shed are still there today. All the lumber is the original except for maybe a few planks that have been replaced.

The fall of 1969 was the last year cattle were shipped by rail. The trucks took over and the railroad stockyards up and down the High Line were either bought or leased by the Counties to be used for shipping cattle by truck transportation. My son Douglas went back East with our shipment in 1969. He had made several trips before. That year he came back in an International pickup from the factory. On the way home he went through Nebraska and had a good visit with Uncle Claude and Delbert Badberg. Uncle Claude filled the pickup box with

Sales Record from Chicago Livestock Commission, 1927

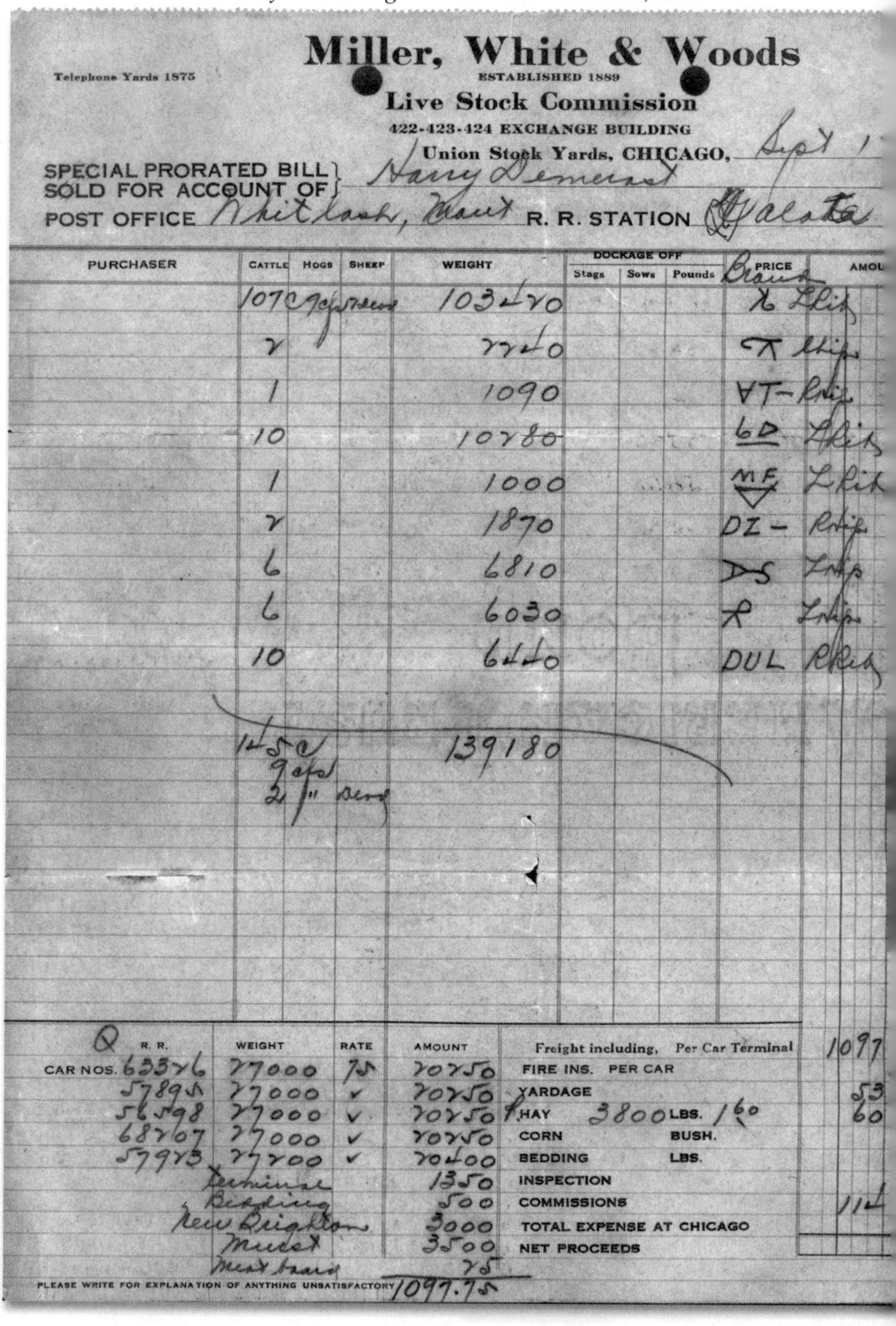

Telephone Yards 1875

Miller, White & Woods

ESTABLISHED 1889

Live Stock Commission

422-423-424 EXCHANGE BUILDING

Union Stock Yards, CHICAGO, Sept 1

SPECIAL PRORATED BILL }
SOLD FOR ACCOUNT OF } Harry Demerast

POST OFFICE Whitlash, Mont R. R. STATION Galata

PURCHASER	CATTLE	HOGS	SHEEP	WEIGHT	DOCKAGE OFF Stags	Sows	Pounds	PRICE / Brand	AMOUNT
	107 C 9 cfs 2 cows			103420				2 L hip	
	2			2240				CT L hip	
	1			1090				VT- R hip	
	10			10280				6D L hip	
	1			1000				MF L hip	
	2			1870				DZ- R hip	
	6			6810				DS L hip	
	6			6030				R L hip	
	10			6440				DUL R hip	
	145 C 9 cfs 2 " dery			139180					

R. R. Q CAR NOS.	WEIGHT	RATE	AMOUNT		
63326	22000	75	20250	Freight including, Per Car Terminal	1097
57895	22000	✓	20250	FIRE INS. PER CAR	
56598	22000	✓	20250	YARDAGE	53
68207	22000	✓	20250	HAY 3800 LBS. 1.60	60
57923	22200	✓	20400	CORN BUSH.	
Terminal			1350	BEDDING LBS.	
Bedding			500	INSPECTION	
New Brighton			3000	COMMISSIONS	114
Must			3500	TOTAL EXPENSE AT CHICAGO	
Meat board			75	NET PROCEEDS	
			1097.75		

PLEASE WRITE FOR EXPLANATION OF ANYTHING UNSATISFACTORY

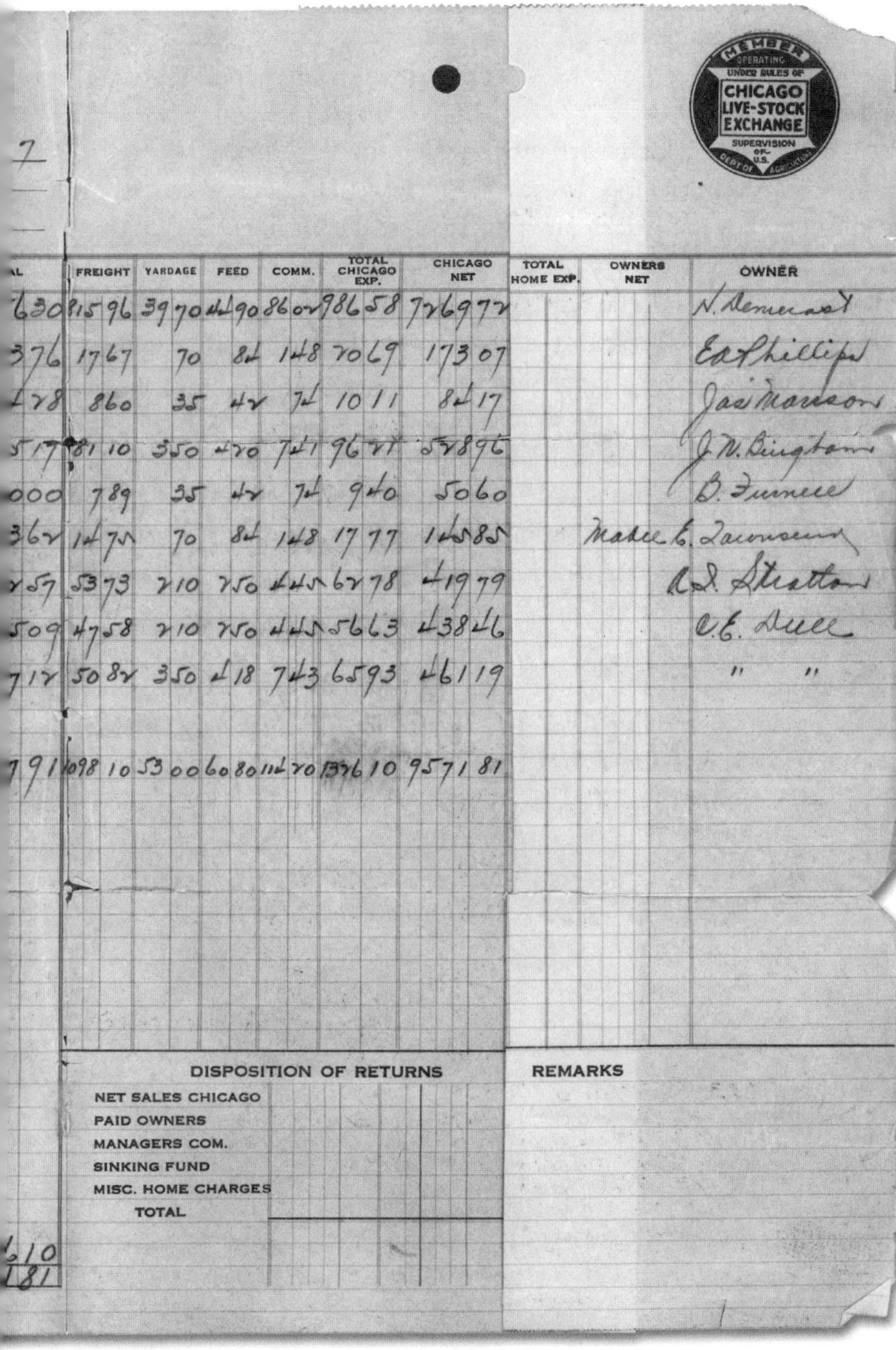

MEMBER OPERATING UNDER RULES OF CHICAGO LIVE-STOCK EXCHANGE SUPERVISION OF U.S. DEPT OF AGRICULTURE

7

…AL	FREIGHT	YARDAGE	FEED	COMM.	TOTAL CHICAGO EXP.	CHICAGO NET	TOTAL HOME EXP.	OWNERS NET	OWNER
630	815 96	39 70	4 90	86 02	986 58	7269 72			N. Demerast
376	17 67	70	84	1 48	20 69	173 07			E. A. Phillips
28	8 60	35	42	74	10 11	84 17			Jas Morrison
517	81 10	3 50	4 20	7 41	96 21	528 96			J. W. Bingham
000	7 89	35	42	74	9 40	50 60			B. Furnell
362	14 75	70	84	1 48	17 77	145 85			Mabel E. Townsend
257	53 73	2 10	2 50	4 45	62 78	419 79			A. J. Stratton
509	47 58	2 10	2 50	4 45	56 63	438 46			O. E. Dell
712	50 82	3 50	4 18	7 43	65 93	461 19			" "
791	1098 10	53 00	60 80	114 20	1376 10	9571 81			

DISPOSITION OF RETURNS

NET SALES CHICAGO
PAID OWNERS
MANAGERS COM.
SINKING FUND
MISC. HOME CHARGES
TOTAL

REMARKS

610
181

a bunch of Osage pine posts. They are crooked as a dog's hind leg, but they never rot in the ground. Osage pines were native to that area in Nebraska.

All the neighbors help each other at shipping time in the fall, gathering sorting and trucking. These days instead of trailing the calves to the railroad, they are trucked to a livestock scale; the ones used in this area are at Fenger's Ranch, McEwen's Ranch or in Chester. From the scales, they are loaded into big "possum belly" trucks and hauled east to the "Corn Patch" states. Order buyers place these cattle to feeders in the "Corn Patch". The cattle are not too many hours on the trucks.

Now with the use of trucks and horse trailers our horses are all trailered to the different ranches to help with the branding and shipping. Pretty soft from the way we used to get our cattle to market. In those days it took four days trailing to the railroad tracks (in Chester) where they were loaded in cattle cars and sent to the Terminal Market in Chicago, Illinois; Saint Paul, Minnesota; or Sioux City, Iowa. We would start trailing the cattle to the railroad on a Sunday and they would be in the market in St. Paul on a Monday, the eighth day. Those years long ago there were no Firestone or Goodyear tire tread marks left on the ole' dirt roads, just "Cow Pies" and "Road Apples"!

Claudie with suitcase and lunch box leaving for St. Paul with cattle—1963

HOME-MADE CATTLE PROD

Around 1925 and 1926 we drove our cattle to Hill, Montana where there was a dipping vat, to dip them for scab. We trailed together with our neighbor Bob Thompson to Hill. This took place after calving and it was a hot job with small calves. What I remember was the oil they used. It was in barrels that had been shipped all the way up from Texas. That was printed on the barrels. Now we got lots of oil wells in parts of our own Sweet Grass Hills.

John Oswood had a home made stock prod, which was attached to a couple of round dry cell batteries. The batteries were carried in old-time canvas nosebags that had a strap attached to go over a horse's head. These bags were actually used to put oats in for the horses. The horse carried the bag with the strap around his neck. When John goosed those cows with his homemade electric prod they really jumped into the vat which was deeper than a cow's body. Then they swam out the other side. They were well dipped in oil and a little wild-eyed. Now days there are lots of electric cattle prods, but back then I'm sure this was John's own invention.

SWEET GRASS HILLS TO CONRAD AND BACK

In the fall of 1933 Schafer and Strode bought a thousand head of calves from Carl Malone at The Flying U Ranch at Choteau. This is the brand . He made delivery on them to the stockyards in Conrad. Carl Iverson and I left home to stay over night in Galata with about ten head of saddle horses. I was riding an iron gray horse when we left home. He was an Adams horse that I was breaking for Schafer, a 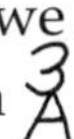 on his jaw and a  on his neck. He was pretty bronky, but about half way to Galata he started getting tired and also sick; and no wonder, all the way he was grabbing his butt and wanting to buck and wanting to get me down with my shadow. So I caught another horse to ride. This happened under a real blind hill, but we couldn't get him to move off the road. We put a lariat rope around his behind and snubbed the rope to our saddle horses, skidded him off in the ditch. Anyhow, about two years later he was picked up running with a bunch of horses near Shelby. In Galata we ate and had a room at the Flage Hotel.

We left Galata and headed out. We two young cowboys knew that Conrad was somewhere southwest. We did have to ride out of our way some to cross the Marias River at a ranch where we were told had a good crossing, as neither of us cowboys could swim. A few miles out of Conrad the black horse that I broke for Schafer got sick; his problem was colic. We knew that for sure. We were now into the irrigation area and there were lanes and pastures, so we put him in a pasture and left him. It was getting dark and we were close to town. Mr. T. P. Strode's Model A. Ford shows up, and he says, "Just follow our tail-lights and we will show you to the stockyards, and if you kids are hungry we might find some chow." Well, it was easy to see our trailblazers had been in town too long and were feeling no pain. Away they went into the darkness and left us with no tail-lights to follow. After some problems we got to the yards. No tail-lights there either. So we found the Conrad Hotel Bar and Café where they said we would be camping for the night. When the waitress brought our groceries and a big steak, Jimmy Morrison asked her what breed of a critter this steak was from. And she said, "An Angus", and Jimmy pushed it away from him and wouldn't touch it. Well, whether it be Angus, Hereford, mutton, goat or jack-rabbit, there were two young cowboys that had ridden some fifty or sixty miles with no cookie wagon along or menu to sort through, who were glad that some cowboy in our crew didn't like Angus beef!

Next morning we drove out to see how the black gelding was doing and we soon found out he wouldn't be helping trail a thousand calves to the Sweet Grass Hills. The six-month old calves got into Conrad next forenoon so they were fed hay in the stockyards and we didn't leave till the next day. These calves had a bar brand about a foot long, called the "Britchen" across their butts just above their hawks.

Ova Brower, Jimmy Morrison, Bill Schafer, and Carl Iverson and I were the cowboys headed for the Sweet Grass Hills the next day. Fritz Sommers had an old truck for our cookie wagon, also oats for the horses and he would set up a tent at noon and also at night. He kept a lantern lit and hanging in the tent all night. That way we could always find our "Home On the Range" when changing guard. He was a real good "biscuit shooter". We hobbled all our extra horses at night. During the day we trailed them along with the cattle.

We had some problems trailing in the lanes until we got out of the irrigated area. With that many calves, some were getting through open gates and downed fences. The owners of these places would get pretty upset and of course always asked who was the Boss of the outfit. Of course Bill was, but we would tell them "that guy with the bright red shirt." That was Jim inside that red shirt, so he got chewed out a plenty. At one place this fellow came on foot, thinking we were going to water the calves in his reservoir. I motioned him to stay back, but on he came and one thousand calves got scared and took out a big bunch of fence. It was quite a mess. At that time the country was pretty isolated and sometimes the cattle would be strung out for a mile. Ova went ahead with maybe fifty head and the rest followed pretty good. Bill and Jim were on the drag. Carl and I each worked a side. We rode a lot of miles and changed saddle horses several times a day.

The first night Jim and I stood guard the first half of the night. From that time on, just one of us at a time took a shift. Just kept them from getting too scattered out. I remember that on this trip the nights were pretty moonlit. That helped us a lot with night-herding those calves. Next camp was next to the Marias River. The following night just before dark we crossed Highway #2 and the Great Northern railroad tracks between Devon and Galata. We camped a few miles beyond there. The fourth night's camp was across Willow Creek on the Devon Road. Our last night was on Strawberry Creek in Harold Henry's' field and it snowed lightly that night. From the time we left the Conrad area until we were in The Sweet Grass Hills, we were in open country. It had all been homesteaded at one time, but most of the people had pulled up stakes and headed for greener pastures. Fritz Sommers chowed us for dinner the last day about four miles south of Whitlash. He was getting a bit crabby and I think fed up with us cowboys and his job. I remember plenty what he said to us. "You so and so's are on your own. I'm headed for the ranch", and he did just that. When we went by the Whitlash School in Earl Thompson's field, Elsie let the pupils stand by the schoolhouse and watch us trail by. And I threw her a little kiss as we went by, and I'm sure she caught it. My little sister, Chick, was going to

Jimmy Morrison, Artie Morrison, Claudie and Johnny Morrison

school there too. But we didn't quite make Schafer's ranch. We put them in a fenced field on the Demarest Ranch.

A few days later after getting them on location, these calves were all heeled and held with a horse and manpower on the front end and branded. There were no cattle chutes in them thar days. Ova, Jim and I did the roping. Calves in those days at six months of age would be around 350 to 400 pounds, not like our 600 to 650 pound calves of today at around six months old.

It was a great trip. Never a dull moment when Ova Brower was around!

A SICK COWBOY

We shipped and trailed together one fall with Robert Sutton. Johnny Morrison was along and Elsie had the Cookie Wagon. The last night we camped in the creek below the Hull Place. Some of us neighbors had built a corral next to a big cutbank, so that took the place of some fence and we did not have to night-herd. So to be good to our chef, I took Elsie in the pickup to town and we had supper. Then Robert and

Johnny went in to eat. When they got back to camp, Johnny was really sick, so Robert and I headed back to town with him. Elsie stayed and kept the campfire in wood. He was really bad, terrible pain and trying to vomit and couldn't. Robert hit a real rough spot on this country road and shook us up pretty bad, and I says, "Robert, you just as well slow down 'cause I don't think Johnny is going to make it anyhow", and I was serious. I had once been told that having a person gulp down all the milk they could hold would make them vomit. We got him into Tommy Smith's eating place and into the little boy's room, and started getting milk into him. Anyhow it did the job. It was really a mess. He maybe had some food poisoning. The next morning he looked a "little green around the gills" and his eyes looked like "two burnt holes in a blanket".

THE TEN DOLLAR BILL

Near the end of one trip to Chester trailing for Bill Schafer, we had the cattle on the rail and heading east. I was pretty happy that I'd get started for the Sweet Grass Hills and not have to stay over night in town, because we were paid $10 for these trips. That's for three days of trailing to the railroad and the fourth day getting home. If I had to stay all night in town, a room and eats and toddy for the body about took my hard-earned ten bucks. Bill says, "Claudie, come around to the car and I'll pay you." While we were walking around the north side of the corral, I spy a ten-dollar bill fluttering in the breeze on a thistle weed. Just as I reached for it, the thought hit me, "Another ten bucks makes twenty and I'm heading for home!" Seeing that ten-spot for the taking was a thrill, but it didn't last for long. Bill reached for it when I did and got it first, and handed it to me and says, " Claudie, here's your ten dollars. We won't have to go to the car." Ranch wages during my growing up years were two dollars a day. Trailing cattle I most always had a bronc on these trips, also a broke horse. The bronc was well broke after a four days trip. I broke a lot of horses for other people for a ten-spot.

BRONC TO THE BAR

One fall when Johnsons and Schafers trailed their beef to Chester, Squeak Johnson, Jim Morrison, Bill Schafer and I were the riders. The last few miles from the yards we got hold of a bottle of Giggle Juice. Being we hadn't had anything to eat since breakfast, we were a little gaunt and drinkin' spirits on an empty belly, you get "loop-legged" pretty bad. Anyhow, we got the cattle corralled, and had to ride through town to where we hoteled our saddle horses. Going down the street our two partners were ahead of us, and Jim said "Lets pass them up!" So we went down the plank sidewalk. I hit an area of sidewalk that was freshly poured cement. We had to pass by a Hotel and Bar, so I decided to ride in and get a little "Toddy for our Body". Well, my partner couldn't get his horse to go through the door, but sorry to say Spot pranced right in! What else could I do, but buy a Round for the House, and have a little nip for myself? I told the bartender I would be back tomorrow and settle up my bill. The bartender also took a little "Cheer" out on the sidewalk to my partner who was still mad at his horse because she wouldn't go into the bar.

The next morning we turned the cattle out to herd them on grass, until the train came to pick them up. Jim and I were with them and out came the Under Sheriff, Jack McDowell, and says we got to go to town and see the Judge, and he asked me if I had rode through the fresh cement. I said I didn't think so, and Jack says, "That sure isn't dried mud on your horse's legs, looks like dried cement to me." "Well, we can't leave these cattle". "No problem, give me your saddle horses and take my car." So we drove the Deputy Sheriff's car to town to see the Judge, and trudged up the stairs into the Judge's Chambers, still chapped and spurred. Judge Shepherd gave us a good fatherly talk, fined us six bucks apiece, and sent us back to relieve the Law and Order who was herding our cattle. I've never been too proud of this episode and a little embarrassed on different occasions. At the time, what was bothering me most was what the new school teacher at Hoffners would think. We had been having a great fun time all summer, going to dances, picnics and rodeos. On the way home I stopped at Hoffners and Elsie thought it was all pretty funny, so that sure made my day. I think the whole county knew about our escapade the next day.

Just a few days later I rode down to Arthur Fey's cow camp on Strawberry Creek, just above where Hales Scalese lives now. Gordon Christian had bought Fey's calves. We corralled the cows and calves the night before and left with the calves next morning for the Prichard Place. It was a long, hard trip and we got to his corral after dark, just Gordon, another fellow and myself. When taking calves from their mothers to trail them some place, you really got to move them on the run and don't let them look back and bawl or you are apt to spill them. After they have gone a few miles they handle pretty good, especially after off their home range. However, their instinct will take them back miles and miles to be with their moms, so when they get to their new home, they have to be kept corralled for a week at least.

The next day I left for home trailing some buck sheep that Gordon had summered for us. It was about 15 miles. Punched beef one day and mutton the next.

TO THE ROCKY MOUNTAIN FRONT

In the fall of 1938 Leo Milles and the Demarest Ranch sold a string of dry cattle to Fred Lewis whose ranch was on Milk River north of Browning, Montana. We were to deliver them in Sunburst. Johnny Morrison and Leo Milles, Jr., and myself were the crew. We put them in a field by the McDermott coal mine the first night. We were a few miles from Sunburst when my horse, Peaches, (who I was breaking for Dave Thomas) throwed a fit and started buckin'. When she was in the air she hit a steer and the boys said we took a couple of end over ends before I got loose from her. It happened in the ditch and the dirt was soft from the roadwork that was being done, which eased my fall some. I had dirt in my ears, nose, mouth and I guess clear to my belly button, and my neck felt like it was broke. Sixty years later my neck still reminds me of that wreck.

We corralled the cattle in the Sunburst stockyards. The next morning Claudie D. could hardly get out of bed and looked like he had been drug through a knothole. That morning we met the cowboy who had ridden from the Lewis Ranch to receive the cattle, leading a horse for one of us to help him on with the drive. Well, Johnny had just gotten over the flu and didn't feel like the trip, and Leo just didn't like to cowboy that well, and for me, I'd seen better days than I felt that morning. Fred Lewis raised and sold polo horses and ranch horses, so I think to myself, "Well, I'll be well mounted", so we turned the cattle out the gate and pointed them toward the Rocky Mountains. To my great disappointment, I soon found out that I was on the most counterfeited "Son-of-a-Horse" and he should have had his head lookin' through a horse collar pulling a plow instead of having that Hamley Riggin' cinched on his back. He wouldn't neck-rein. I had to G-Haw him to get him turned around and he was real stub-toed. We left with lunches that had been made that morning at the Cafe tied behind our saddles and a fifth of whiskey. Dick said he didn't know who he was going to get to help him trail, and this counterfeit was the only gentle thing on the ranch. How I did wish for my own horse, but she was already headed back for home. After the way she did me up the day before, she deserved that reservation trip and a few more days carrying a wet saddle blanket

We didn't get too far that day, as the cattle were hungry. We stayed with a bachelor who had a field for the cattle and he treated us great. The next day put us into the Blackfoot Reservation. At that time it was all grass, no farming. It was another one of those dry runs. No biscuit wagon so we were on our own. Coming over, Dick stayed with a crew on the north side of Land Slide Butte. They were building fence. We trailed all that day and all night and at daylight came to this camp. They were just eating chow. We were a bit hollow and dry, and I'm sure my belly thought my throat had been cut. But the cattle did great picking grass as we trailed. And I guess that's the main thing, cattle need to get fat and shiny, cowboys don't want to get fat anyhow. On leaving the camp our belts were let out a couple of notches and we were ready for another day. That night we stopped at a ranch on Milk River, the Johnson Ranch. The last day was just a half a day's drive to the Fred Lewis Ranch. Dick brought me home the next day.

CANADIAN CALVES BORDER CROSSINGS

Our ranch is about six miles from the Canadian border, so many of our friends and neighbors are Canadians. We frequently bought Canadian calves and even though they were pretty close to home, trailing them home amounted to more than just that; we had to go through the Canadian and United States Customs.

For several years Dad and our family bought calves from Herb Samons on Deer Creek in Alberta, next to the border on the north edge of West Butte. Dad and I were at his ranch to look at his calves. On getting ready to leave, Dad got in the car. I didn't, 'cause Herb really liked to visit, and I knew we would be there for awhile longer. While Herb was talking to Dad, he kept trying to get his foot on the running board of the car. This was about the time that running boards were out-dated and this car didn't have one. Finally he glances down at his foot and says, "Harry, you broke the darn thing off, eh?" But he never stopped visiting.

We purchased calves from other different Alberta neighbors, and trailed them home. I would ride and lead a couple extra mounts and enter into Canada through our Canadian Port six miles north of Whitlash, but when we headed for home with the calves, the customs let us cut across by Albert Feys and on home. It was a pretty long drive. The Canadian Government had a good set of corrals and a scale at the Old Whitlash townsite, and the calves would be weighed there for duty. Things are stricter now. No cattle are brought in or weighed for duty these days except at the Coutts / Sweet Grass Port of Entry, which is thirty-five to forty miles directly west of the Whitlash Port of Entry. Those corrals and scales were built when George Ross, an Alberta rancher, trailed cattle across to the Arthur Fey Ranch that he had bought. A Canadian official would come from Coutts Port of Entry to weigh them at the Whitlash scales for the duty. In the fall the cattle were shipped to our terminal market either in St. Paul, Sioux City or Chicago. When World War II was on, all Ports of Entry were closed to livestock and the scales and yards at Whitlash were sold. Robert Sutton bought the scales. After the war livestock once again began coming across the border through the Coutts Port of Entry.

CHAMBERS RANCH COWS AND CALVES

Bill Schafer took me down to the Harry Chambers Ranch on the Marias River south of Shelby, to help him gather cows and calves as he had bought the calves. Harry and I would leave the barn at the Chambers ranch before daylight and it would be after dark on getting back. One day we were just a little south of where the fairgrounds are now, pretty close to Shelby. We rode across a nice looking little stream of water and I was really thirsty, so I gets off and takes on a big drink. Boy, was it terrible! I'm thinking, "I have drank alkali, soda, and sulfur water, and been forced to drink out of a few cow tracks in the mud, and have had some pretty bad moonshine whiskey", and on catching up to Harry, says to him, "that is the worst water I ever drank!" He just chuckled a little and says, "Well, you just drank out of the Shelby sewer", and he gets out a big sack of chewing tobacco and says, "take a fist-full of this." With the chewing tobacco and the bad water, my complexion I'm sure wasn't too rosy for several miles. We got the cattle finally gathered, so it was three days before I got to see what the outside of his house and barn looked like in daylight.

For the next few years, Bill Schafer bought his calves and we trailed them to the Sweet Grass Hills. One year Harry Chambers got grasshoppered out and he trailed cows and calves to Bill's. I wasn't on that drive, but that fall helped Chambers trail the cows home. No calves, left them at Bill's. The first night we put them in a field at Nels Sveum's, next to Grassy Butte. We stayed with Nels. We had no Biscuit Wagon along, so I took on a big breakfast. When we crossed Willow Creek the cows got a drink there. By night they were getting dry, but they knew they were headed home and would get a drink of water when they got to the Marias River. When night caught up to us, those cows just kept a-goin'; a little reflection in the sky of Shelby lights helped us keep them pointed. He had a few white roan cows, and this doesn't seem right; but a roan critter or white horse just doesn't show up in the dark, but dark colored animals do, especially in semi-darkness. Just past daylight, we came to a ranch and they were eating breakfast. They lived a few miles from the railroad track and Highway #2. So that was all day and night and no biscuit or drink. Only had Big Sky Montana for filling and the Milky Way for moisture. These cows carried the ∃X on their ribs. I was riding a real good strawberry roan horse that belonged to his ranch. Bill Schafer bought calves for several years from Harry and I was on all those trips from the Marias River to the Sweet Grass Hills. I trailed the saddle horses to Chambers and helped trail the calves back, more than sixty miles each way.

Harry Wehr needed a horse for his garbage wagon in Shelby. This Speed horse that I used to ride had gotten a little stove up, so I had broke him to work for a mowing machine team. On this trip, I led him along. Harry Wehr used him that winter. I forget now the details on getting him home again.

One year Oscar Iverson bought fifty cows with the ∃X brand. Edgar Woods and I trailed them from the Marias River to the Sweet Grass Hills. Oscar was along with the Biscuit Wagon, and "By Yeasus", he was a good camp-tender. The last night on Strawberry Creek, Oscar left our bedroll and some grub and went on home. We put the cows in a pasture. It looked like snow to us, so we loaded our grub and bedroll on our saddle horses and led our horses about a mile across country where I knew was a barn to get into for the night. And a good thing too, 'cause it snowed and got pretty cold.

When Chambers sold his ranch to Melvin Underdahl, my Dad and I bought his cows, around a hundred head. He delivered them to the stockyards in Shelby, but we sold them to Pop Dennison for a profit and didn't have to trail them to the Sweet Grass Hills.

COW TRAILIN' TALES

For many years the only way we had to move cattle when buying or selling, was by trailing them with horses. I rode many thousands of miles horseback punchin' cows and calves over the years. It was not until later years that trucks were available for transporting cattle. Now, in the year 2001 we still move our cattle around to different parts of our ranch with horses, but for any longer hauls and shipping, we have our own trucks.

We bought fifty head of cows and calves from Johnny Brinkman one fall. I left home leading an extra saddle horse and stayed at the Hawk's place on the way. The next day I rode on and crossed the bridge at Meisners, and then on down the Marias River to his ranch. Dad brought Scott Morrison down the next day to help trail the cattle home. About a week later we got around fifty pairs from Charlie Brinkman. If we had known both bunches were for sale, we could have saved a few riding miles. He did deliver his cows to the yards at Chester. They both had to cut down on numbers because the grasshoppers had hit them real hard. Grasshoppers like grass just like cows do.

Henry Kolstad leased some grass from Ben Sirgenson. (Ben had bought the Robinson place and later sold it to Alex Lakey). Matt Furnell and I trailed the cows and calves to the Hills. It was a slow drive as the calves were pretty young. We rode two of his saddle horses. I remember Alan Kolstad was just a little guy with freckles and red hair crawling around on "all fours"! We stayed at their place the first night, leaving the next morning and crossed Highway #2 and the Great Northern tracks at Tiber crossing.

Another fall we bought the Prescott heifer calves and put them in Bud McDowell's corrals the first night. Just before getting to Bud's we hit a real wet snow flurry, and I just could hardly see. At the time I wore glasses and was afraid to stick them in my pocket as they might get broken. So I found a gopher hole and put them there, hoping the next morning I could find them. The next day they were there all right, took a little lookin' though, and we made it on home. *(Now I'm wondering why you thought your glasses were safer with a gopher than in your pocket?—The Editor)* Red Morris had bought the steer calves and we trailed them together. Jim Morrison, Irvin Brown, Jim Woods, Dad and I were the cowboys.

We bought Prescott's yearling steers again another year and Dad and I trailed them home. Other calves we bought and trailed home were from Dot and Red Morris, Melvin Madison, Mrs. George Murray, and Howard O'Loughlin. About 1950 people started trucking cattle. Walt Clark, Oswald Bradley, Green Buffington, Jack Griffith, Jack McDowell, and a rancher up by Medicine Hat, Alberta, and a rancher by Formost, Alberta delivered calves by truck that we had purchased from them.

Elsie cookin' dinner

YOUNG COWHANDS ON THE TRAIL

One year we took our kids and a couple of other kids trailing cattle to Chester. Elsie was driving our Cookie Wagon. Doug, Auverne, Robert Todd and Bruce Todd were the young cowboys. Dennis Rapp who worked on the ranch and myself were the rest of the crew. Robert and Bruce went the first two days and then went home. We had to night herd the last night out of town. It was kind of chilly and we had a tarp over the top of the stockrack of the pickup, and Doug and Auverne bedded down in there. We had the cattle surrounded with a fence on three sides, but the open side one of us had to keep watch. I tried to get Elsie to sack-out in the pickup, but she just kept a big fire going till morning, sitting there huddled over the campfire, just like a squaw outside her teepee. She was sure going to take care of

Doug on Pronto, Bruce on Robley, and Robert on Tess

us night-herders! It even spit a little snow during the night. It was petty cold and frosty that morning so Elsie, and our young cowgirl and cowboy went to town for breakfast. We have some real good pictures of those young cowhands on that trip; Doug on Pronto, Bruce on Robley, Robert on Tess and Auverne on a horse that belonged to Roy Todd, and a picture of Elsie at the pickup getting dinner ready for the bunch.

VI / Ranch Work — Then & Now

THE SHEEP BUSINESS

In the fall of 1926 we took a 1000 ewes on a share deal of the lambs and wool. They came by rail from the Highwood area. A sheepherder and his dog came in the caboose; his name was Fred Bethune, and the dog was jet black by the name of Tucker. Fred herded for us that winter and helped us lamb that spring. When he left after lambing we kept Tucker, and we had him for years. He was a good winter worker, but when summer came, he just didn't like the heat. So we got a longhaired female dog from Harold Henry. We had her still, when we sold the band in the fall of 1935. While we had this band and needed a lambing crew, we went to town to get help. In those days the streets were full of guys wanting work. On approaching one of these guys looking for work, the rancher would ask the fellow, "Have you had your tongue split?" If he was a seasoned lamber he would say "Oh yes!" If you asked a fellow that question who hadn't lambed sheep, his answer would be, "What do you mean, 'tongue split'?" (Answer), "So you can lick both sides of the newborn lamb at one lick and at the same time." (Mine has already growed back together).

I always was the night man and liked that job 'cause your work was in one spot all night. When a ewe lambed she and lamb were put in a little small pen, called the "jug" and suckled and their navel was iodined to keep away infection. One fall when we sold and shipped our lambs, we weighed them and a lot of other sheepmen's lambs at the Joe Berthelote Ranch on Miners Coulee. Jim and Norma Woods live there now. Another fellow and I *on foot* and with our dogs started to trail 3000 lambs to Devon to be loaded on the rail. Without a doubt that was the worst trip I ever took with livestock. It was hot and dusty and my dog wouldn't help me on account of the heat and prickley cactus. About the second day I was so galded my steps were getting pretty short and getting more spraddled legged by the mile. Oh, for some Bag Balm or even gun grease! On the 4th day we had them on the rail. No more trailing sheep without my saddle horse for this cowboy (mutton puncher)!

Years ago Dad had a band of sheep four or five miles south of Hay Stack Butte on summer pasture rented from Dick Wickum. As a little guy I went with Dad to take groceries to the herder. We went by team and buggy. After arriving at the sheepherder's sheep wagon (a home on wheels), Dad fixed me a sandwich and parked me on the wagon tongue. The herder had a pet gopher and Mr. Gopher hopped up on the wagon tongue, then into my lap and started helping himself to my sandwich. Dad came to my rescue. I must not have been very old, because I remember the gopher, but nothing else of the trip.

When my sisters and I were kids we bottle fed the "bum lambs", which were the extras from sets of twins that their mom's couldn't take care of. When these lambs were sold we got the money which we saved for a couple of years. My sisters pooled their money together and bought a piano. I bought my first saddle with my share.

Helen Demarest, Edith Morgan Claudie with lamb and Edna Morgan

CALVING IN THE EARLY NORTHERN SPRING

Calving season is most always cold and spring blizzards are common during March and April in Northern Montana. We don't leave the ranch for any length of time during calving season. When a storm is coming, we get the cows that are ready to calve soon into a shed for protection from the storm. If the weather is good, they stay out in the calving pasture. We check on them every few hours, day and night no matter what the weather.

We have saved many a calf by being there at the right time. Sometimes calves are lost for various reasons. They could be coming backwards and need to be pulled, or frequently they are born beside a pond of water or creek, and the calf flops into the water and drowns right after he has been born. This year (2001) that is not as much of a problem, because of the drought. Usually every little buffalo wallow is filled with water.

In years past we had Hereford cows. A big problem was that when there was snow on the ground, their bags would get "sunburned" and they wouldn't let their calves have any dinner. So we would have to run the cows through the chute, maybe forty to fifty head, and grease up their bags with Bag Balm. If the calf was really hungry we would tie the cow's legs to keep her from kicking, and put the calf to her while she was in the chute. We had to milk out the cows whose bags were too full. These cows with their raw and sore teats would keep lying down, keeping the calves from getting any dinner. After years of that problem, we finally switched to Angus cows. They have black bags and teats and do not sunburn. However, a black white-faced cross breed with a white bag sunburns just like the Herefords.

Calves being born hind feet coming first instead of their front feet and head, have to be pulled from the cow, and taken as fast as possible. The navel breaks loose and until his head is out, the calf gets no air through his nose. But with a good "calf-puller" made for just such a situation, it works pretty fast. So at calving time you better be a damn good cowboy and also a vet.

When a calf is born dead or dies, we take the hide and tie it on a twin calf if we are lucky enough to have a set of twins. It is tied on tight and when the mother cow of the dead calf smells 'the scent' of her dead calf's hide, she thinks it her own and is a Happy Mother Cow. The "jacket" is only left on for about twelve hours. I have never seen this transplant fail; we have done this many times.

Sometimes it happens to be a nice and warm day with the wind blowing about sixty-miles-an-hour. You know, the kind of wind that about pulls the fillings out of your choppers if you don't keep your mouth closed. If a cow tries to calve with her rear-end facing into the sixty-mile-an-hour gale, she just is not going to get her new baby into this world. She has got to get up and change directions. Now, that is pretty windy, huh?

HEEL FLIES

Before we had pickups with stockracks and horsetrailers, we used to ride from ten to fifteen miles to different neighbors to help them at calf branding time. I have never missed a neighbor's branding since in my middle teens up to the present, and during that time the herds had to be gathered and corralled before the sun got things too hot. The reason was a little fly called "The Heel Fly". It went to biting the heels of the cattle and they went plum crazy, and stuck their tails up in the air. No way could a bunch of cowboys corral them once the heel flies got to them. If there were lakes nearby, that where the cattle went until the water was up above their legs. If there was no water, they headed into the brush. Dad used to say, "Got to get them corralled before 'flag-day' starts!"

I have always said, "One of the greatest improvements in the live-stock business is when a pour-on medicine was made to put on their backs to take care of the heel-fly problem". At present cattle can be corralled or trailed at any time of the day, hot or cold.

SPRING BRANDING

When I was a young teenager, Bert Furnell got me started roping and heeling calves. We were branding up at Bob Thompson's and Bert Says to me, "Get on your horse and heel these calves. Let the horse do the hard work". For several years I was the only one in our area who could heel calves. A few years later Jimmy Morrison, Troy Lakey and Irvin

Brown started heeling at all the brandings in our area. All calf heelers used to "tie hard and fast", (rope tied solid to the saddlehorn) but now we all "dally" (a couple of wraps around the horn). A long time ago at Walt Clark's I was heeling calves "hard and fast". I was going back into the herd, and my horse started to buck. Howard O'Loughlin, who was one of the ground crew, stepped into a loop of the rope dragging on the ground. As he was starting to make a little 'altitude", Walt cut the rope. While this was happening my horse ran his head into the side of a barn, and I got him under control. From then on I was a "dally" roper. Doug also learned to heel calves when he was very young, and yet today ropes at all the neighbor's brandings.

In our community every May and sometimes into June, all the neighbors get together and help each other on branding day. Early in the month each ranch picks their "branding date". We usually brand sometime during the last ten days of May. Early in the morning cowboys arrive at the branding corral with their trucks hauling their horses to help Doug and I round up. By the time the cows and calves have been gathered, the rest of the branding crew is at the corrals, and the fire is going, heating up the branding irons. These days we use propane for the fire in a portable branding stove, called a pott.

Each year the branding crew members change some, but usually Buster Brown, Dave McEwen, Helen Brown, and sometimes Dave Bryan are the ropers who heel the calves, pulling them out of the herd and up close to the branding fires to a set of wrestlers. Calf wrestlers often are Warren Iverson, Bobby Thompson, and the younger Thompson boys, Chance, Jeff and Nick, some of the hired hands from different ranches, often younger guys who are there for the event, and sometimes even Helen Brown. She likes to give up her vaccine gun or ropin' and get a little manure on her Levi's! Toni Wehr, Helen Brown, Crystal Demarest or Donna Bryan usually are the vaccinators. Ron Wehr gets the job of checking for "nubin" horns which if found are removed by a gadget made for that purpose. Years ago Matt Furnell did the calf cutting at all of the brandings. When Matt sold his ranch, then I took over the cutting for years. Now Brian Ratzburg does the castrating of the bull calves at our branding. He sort of took over my job, as he has a special tool, which does the job without cutting the calf. Before that tool was available, I castrated all of them using a knife. The new way is much less traumatic to the calves. Earl Thompson, Albert Fey and Doug and I usually did most of the branding. Irvin Brown keeps the "tally". Jim Sutton, Lon Hoppe, Will Clark, Mark Krupar, and Sid House are often part of the crew. Someone has to pick out the correct irons for each calf. We now use the , on all our calves, and the 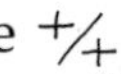 on Auverne's and Tim's. We used to also brand many of them with the . All this goes on amidst a corral of milling cows and calves all bawling at once with the wind blowing the branding smoke every which way. Many times there are visitors who come to watch, take pictures, and join us for the day.

Brandings turn out to be great get-togethers. When the branding is finished and the cows and calves are turned out of the corrals back into the pasture, the crew shares in a little beer, pop, or Black Velvet before going back to the ranch for a great feast. These feasts are prepared by the Best Cooks in the Sweet Grass Hills! All of them should have a Gold Star for their efforts! Our cooks, my daughter Auverne, and granddaughters Paula and Crystal, prepare roast beef, ham, sometimes turkey, mashed potatoes and gravy, baked beans, vegetables, maybe a casserole or two, a variety of salads, rolls, and for dessert pie and ice cream. No matter what your favorite kind of pie, it's there on the table just for the takin'. There are usually a couple of neighbor ladies who come and help our cooks on the morning of branding. Gloria Fey and Jeanette Brown are there most every year. Doug and friends have hosed out the dried winter mud clumps and dirt from the big garage and it gets its yearly clean up, because that is where the buffet feast is held. Tables and chairs are set up for thirty to thirty-five people.

Doug and I attend three or four of these brandings a week in the spring, until all of the ranchers have branded their calves. We work hard, show off our skills a bit, get the job done, enjoy a great meal, then sit around and lie and brag the rest of the day. I have heard some say that these are really "cowboy parties". Everyone—the cowboys, cowgirls, kids, cooks and visitors seem to have a really good time. Last year Dean Hellinger took a great video before and during the branding, and of the "chow table" and presented us with some copies. Thanks, Dean!

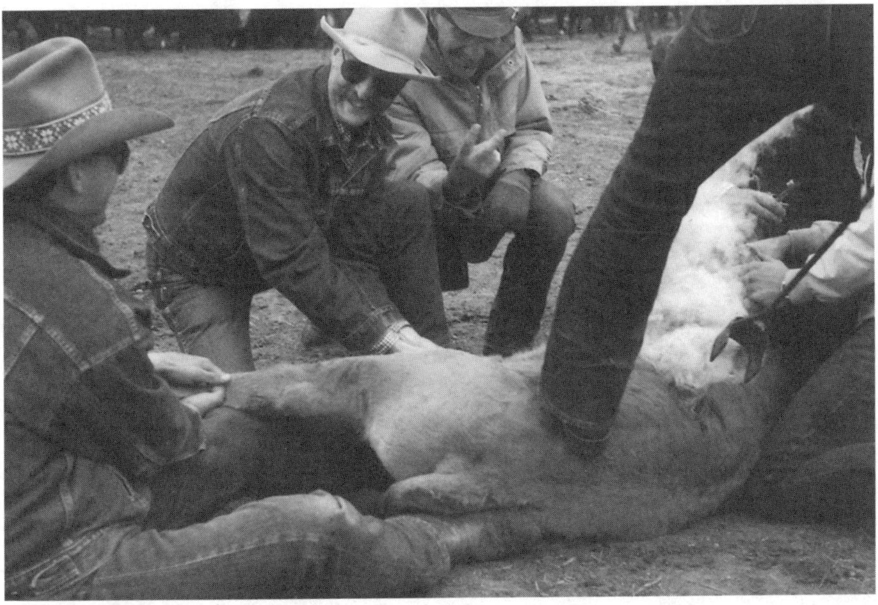

Glen Thompson, Claude Ritter and Claudie

Ron Wehr, Warren Iverson, George Mader, Earl Thompson, Helen Brown, & Doug

Doug, Claudie, Dennis Rapp —1940's

Demarest calves

Brower Ranch in Canada, Claudie, Tim Dieffenbach and Doug—1959

BRANDING WITH OUR CANADIAN NEIGHBORS

When Doug was a teenager, he and I went over to the Brower Ranch in Canada every year at branding time and helped them brand the "Walking Cane Brand" . Bill Brower and Ken changed off heeling, there were three or four sets of calf wrestlers, and we two Yankees did the branding of about 300 calves. Doug and I used their saddle horses when we gathered each day for the first bunch. The next day we moved saddle horses and equipment farther east to a cow camp north of the old Spencer Ranch on Milk River. These calves belonged to Henry Wertheimer who leased this ranch and the Brower family managed it for him. Around 1500 calves were worked here. In a matter of several days these calves branded –N–.I stamped the N and Doug did the bars. One day Wertheimer's son was there from St. Paul taking in the "going ons" and that night he says to me, "I watched you close all day long and figured for sure you would make a mistake and get the N upside down." We kidded each other later that day about his remark. I could have made a Z real easy, but you can't get an N upside down. But if the iron handle were welded on the wrong side of the N it sure would have been backwards!

Henry Wertheimer lived in South Saint Paul, Minnesota and I think that more cattle passed through his ownership than any man in the United States. He had several ranches and feed lots. He was an every day buyer in the Main Alleys in the South Saint Paul Stockyards. For years he was always riding a "Clay Bank" colored saddle horse. Charlie Govern was the owner of the South Saint Paul Commission Co. When Henry riding the Clay Bank horse and Charlie on Palie, his palomino saddle horse, got together dickering on a string of feeder cattle, a lot of cartwheels were passing through their hands.

Not far from the corrals was a lake and lots of sea gulls around it and the Brower kids and their cousin were using calf nuts, "Rocky Mountain Oysters", as bait on a fish hook and line. Mr. Gull takes on a little dinner, takes to the air, then they would pull him back from out of the sky, still having their bait and a big bird besides—great sport and good fishin'. At present Warren Brower and family are on the ranch in Canada. Warren was one of the cowboy kids who were pulling those gulls out of the sky.

Brands have interested me ever since I can remember. I was a "brand crazy" little kid who was always walking around everyone's horse to look at his brand. And these were the days when everyone used horses for transportation. You will notice on the photos of my cardboard ponies, each was branded. Some had had "previous owners" and were rebranded. The  on some of the cardboard ponies is a brand I made up when I was a little kid.

The *State of Montana Brand Book* had all the registered brands listed and being used from 1873 to 1900. There are many Sweet Grass Hills ranchers' brands listed. Another reference I have is the 1903 State of Montana Brand Book showing the brands being used for that one year. I also have two other large brand books from Montana listing all of the brands and their owners. Those books are well worn. A tradition of mine is to add their brand to the address of everyone to whom I send Christmas cards. I've done a lot of looking-up in those books! My return address includes the brands of the Demarest Ranch.

My Dad's brand was the . He brought it from the Wallace Porter family from Great Falls who had homesteaded on Breed Creed between the Morgan place and the Brown Ranch. Mrs. William King, Sr. was a Porter. The William Kings made their home on Breed Creek just across the border in Canada.

My brand is the . I bought it directly from the Montana Brand Office about 1950. Another brand I own is the . It belonged to my long-time friend, Jim Morrison, and I bought it at the auction sale following his death. The  originally belonged to my Grandpa John Brown, Mom's dad. When they left the Brown Ranch and went back to Great Falls, Dad purchased it and later it belonged to Auverne and Doug when they were kids. Now it belongs just to Doug. We have the original  branding iron that Grandpa Brown made of copper when he worked for Anaconda Copper in Great Falls.

I bought the 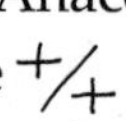 from a rancher at Forseth, Montana and gave it to Auverne and her husband Tim. The 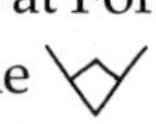 was never re-recorded after Matt Morgan's death, so I purchased it from the Montana Brand Office in Helena. That brand now belongs to my granddaughter, Crystal and her husband, Matt.

When we brand calves in the spring, we have "a lot of irons in the fire" as we are using several of these brands belonging to our family members. Auverne and Crystal are always out there checkin' to see that their calves don't get branded with the wrong brand!

When Thomas Francis Morgan sold his ranch to the Browns, Dad bought three unbroke horses from him. They had FM on the left shoulder. One sorrel mare with a flaxen mane we worked for many years. The two black ones I broke to ride. Tom also branded a 50 on the thigh. These brands were never owned by us, but I remember them from these horses and because the Morgans were close neighbors.

HAZARDS OF BRANDING DAY

Years ago Doug and I went over to Jim Wood's to help brand. We were out riding and saw Jim's hired man after some cattle on a flat with some pretty tall sagebrush. We saw his horse fall with him, so we got off our horses and hobbled them a little ways back, before walking up to the hired man's horse that was on the ground. The horse was lying stretched out and as we came up he did raise his head up, but never moved otherwise. Both of the rider's legs were pinned down about to his knees under the horse's body, next to the horse's belly. The man's head was just half way between this big ole' white horse's hind feet and front feet. He wasn't hurting too much and talked to us. We were afraid if we made the horse get up, those four feet so close to the cowboy's head would kill him for sure. Just then Gordon Christian drove up in his jeep. He cut the latigo on the saddle, so in case the rider's foot was caught in the under side stirrup, he would not be drug if the horse did get up. What we did was to put a lariat rope around both of the horse's hind feet, and the same with the front feet. Then the three of us rolled the horse off the rider's legs. Believe it or not, that cowboy crawled away from that horse just like a snake and soon was standing on his own feet beside us, and the horse got to his feet like nothing had ever happened. It's always been a mystery to me how horse and rider got in that position. The guy and his saddle left with Gordon in his jeep and we led his horse, and started gathering cattle for the branding.

Doug got the job of vaccinating with the vaccine gun, and somehow stuck the needle into himself. A little later Linda Berthelote who was parked on the top pole of the corral fell off and crippled herself up. I stayed healthy all day. Most brandings are "fun days" without all these mishaps. It was a good thing we were only "Two Whoops and A Holler" from the cowboy and saw him and his horse disappear into the sagebrush.

HAYING – 1920'S TO 2001

For a ranch just the size of ours, it took about five people to get the haying done. I was about fourteen years old when my Dad handed me the lines from a team of sorrel geldings and says, "Get on this mowing machine and start cutting hay". The mowing machine was a John Deere No. 5 with a five and half foot sickle bar. Every night I had to sharpen sickles. I sat on a seat and pedaled to turn the grindstone around which went through a casing filled with water to keep it wet. The sickle bar was held in position by hand and two sides of each section were sharpened. Later I used a small stone with a handle to sharpen the sickles. Now our modern equipment has a sickle that stays sharp for the whole season.

For quite a few years Mabel Mae, my sister, raked the hay into windrows, then after a field was finished she went down the rows of hay and bunched it into piles. We always had a man and a team with hayrack pitching the hay on by hand. But we did have a system of laying a big heavy rope on the bottom of the floor of the hayrack and rolling the hay out the back-end, up on the stack. This is kind of hard to explain how it worked but it saved a lot of back labor getting the hay unloaded.

After a few years we started using teams on bullrakes which pushed the piles of loose hay onto the overshot stacker. The overshot stacker had long wooden teeth laying flat on the ground. The hay was lifted up on the wooden teeth of the overshot stacker, powered by a team pulling a cable. The load of hay was dropped off the stacker onto the haystack where Dad placed the piles with a hayfork to the outer edges. This was hard work for him, and he had to use a tall ladder to get up on the stack as it got higher. The team hooked to the cable was then backed up to be ready for another load. The bullrakes and the overshot stacker were big labor savers compared to the first method I described. We always had to have some extra hired hands during haying season.

My job was getting up very early to wrangle the horses, getting the work teams into the corral. The teams would always be caught and harnessed before breakfast, which was at six o'clock. My Dad most always cooked breakfast. "Come and get it or I'll throw it out!" was his call. We had a big ranch bell that was always rung to call everyone in for meals, too. When we were haying at the Hick's place, lots of mornings I was late for breakfast, because of the time it took getting the teams together off that rough country and down to the corral.

A few years later tractors came into use. We got a Model B John Deere tractor and a tractor mower and a John Deere baler. The baler had a motor and belt, with a bale chute that came out the right hand side of the baler instead of behind like they all are now. At that time we hauled the bales with a team. The bales were loaded by hand onto the hayrack or wagon, using a bale hook in one hand to help pick up it up. We had a conveyer that took the bales up onto the stack. One person put the bales onto the conveyer and the other person on was on top of the stack. Next came a pop-up loader hooked onto the side of the hayrack. The driver drove so the bale was picked up off the ground and went up a chute. A man was on the hayrack to place the bales. This was a real labor saver.

We also bought a John Deere Crawler and pulled the baler with that which was great in rough steep country. While trying to unthread a burr on an oil line in a tight spot on the 1951 John Deere Crawler tractor, the ring on my finger touched the frame of the tractor and also the battery post. It arced and burned the ring apart and my finger's flesh to the bone. Boy, oh, boy, talk about smoky hot flesh! It so happened that at one time I had read if you got an electrical burn and were around tractors, to take the oil dipstick and smear on the burn. So after my finger quit smokin' that just what I did. I went to the doctor later.

We were the first in our area to buy a self-propelled swather, a Hesston 500. It didn't have a cab or air conditioner. We now have a

Hesston with cab and air conditioner. We are on our fourth John Deere baler, a 346-power take-off that we use yet today. Most of our neighbors put up big round bales.

So these last bunch of years the crew is two, Doug and myself. I still do the baling. Doug does the swathing and picks up and stacks the bales with a self-propelled New Holland hay wagon that holds 119 bales. It picks them up off the ground and stacks them on the bed of the wagon and automatically unloads the stack on the ground. The bales never have to lie overnight in the hayfield. By evening they are all stacked. In the days previous to this, if the baling got ahead of the stacking, the bales sometimes got rained on. Then they all had to be turned by hand, sometimes several times, in order to dry them out before they could be picked up and stacked. One year, my grandkids will remember this, it rained on the bales at the Hicks Place. Many of the bales were on pretty steep hillsides and SanDee, Craig, Brian and their folks all helped turn them. Then we had more rain before they dried, and we had to turn them all over again.

There are also hay machines that makes big round or square bales that weigh 1000 to 2000 pounds. Special equipment is used to handle them when feeding livestock.

Progress and times have made a lot of changes just in my lifetime.

Hand loading bales, Doug, Ronnie Stonechild, Ted Kibble—1956

MAYRATH

RYE HAY

It was in the late 1930's or 40's that a rancher, Billy Welch, who lived between Cut Bank and Browning, bought about thirty head of two year old heifers from us. He and Dad made the purchase deal over the phone. It was a "Sight-Unseen—Gentlemen's Word" agreement; he had never seen these heifers. It was March, and the cows were to calve in April and May. Billy rode horseback from his ranch to ours to receive them. During that evening he and Dad had a great visit, and the conversation turned to using different kinds of hay for feeding. When rye hay was discussed, Billy said, "It's about as worthless as a snowbank." Dad never said anything pro or con about rye. The next morning on looking at his newly purchased heifers, Billy remarked, "What in the world have you been wintering these cattle on? They are really fat!" Dad responded, "Oh just that rye hay that you "snow-banked" last night!"

Dad, Auverne and Claudie—Rye field 1940

Billy started out heading toward the Rocky Mountain front with the heifers, happy with his purchase. I rode along helping him trail for about twelve miles. My Mother made him take some sandwiches. Where his stops would be didn't seem to worry him much. He did have a pretty good-sized bundle tied on the back of his saddle, so he was probably well prepared for his trip home.

HAYING AND FEEDING CATTLE AT THE HICKS PLACE

The Hick's Place is five miles from the ranch and located in the foothills of East Butte. We put up a lot of meadow hay, brome, timothy, red clover, nut grass, blue pod and vetch. Besides haying the meadow, we hay the upland ground hay and it is pretty darn steep for machinery. When we used to put up loose hay, it was stacked in four different stack yards. We kept our cattle up there two to three months in the winter and went up every day to feed. We fed with a team and a bobsled. Later on when we got a baler we used just one single horse on a platform with wooden runners. It was easy to load bales on and I could feed several hundred cows in a hurry, of course taking a number of trips. However, we fed pretty close to the stack yard. I broke a lot of horses to ride on the rides up there to feed, and it was an every day trip, deep snow, shine or blizzard. When Doug was home from high school on weekends, or on vacation, he liked to stay at the cabin up there and batch and do the feeding. Cooking and heating were from an old coal and wood cook stove, and an old kerosene lamp or Coleman gasoline lantern provided light in the evenings. We had a dry cell battery radio to listen to. He really looked forward to those days!

Now we have our own self-propelled hay wagon, which hauls 119 bales and we haul all the bales from the Hick's Place to the home ranch.

Feeding loose hay at Hick's Place 1931

Before we got our hay wagon, Ron Wehr was doing custom haying, and for a couple of years he did the hauling for us. I said to Ron. "I don't think we can haul out of there with a hay wagon because it is too steep and rough". All he said was, "We will see!" He hauled bales for us that season from the Hick's Place to the main ranch.

These days we take the cattle up there in late summer to graze, bringing them home in October for calf shipping, then returning the cows up there for a couple more months to graze. We again bring them home in late December and start feeding them bales of hay for the rest of the winter.

Arlene Demarest Ritter, Susan Ritter, Claudie, Uncle Claude, Dad, and Frank Demarest at Hick's Place

Hick's Place haying cabin, Elsie and Claudie 1957

CUTTING POLES IN EAST BUTTE

We used to cut timber in our upper mountain field for corral poles and shed material. Most all of our neighbors and Canadian neighbors hauled a lot of poles from East Butte. They were Jack Pine and Lodge Pole Pine, from four to six inches in diameter and fourteen to sixteen feet long. The lifespan of these poles wasn't too long, but if you peeled the bark off them, they lasted longer. We used an ax to cut them down. Sometimes I was alone, and other times there were two of us. This was in the 1930's and 1940's.

We would take the front running gears of the wagon with the reach included and drive the team hooked to it and climb a long steep hill to where we would cut fifty poles. That was a good load with the big end of the poles loaded on and the small ends would drag on the ground. Those ends dragging worked as a brake on the long and very steep hill to the bottom. The poles were chained securely to the wagon. When we reached the bottom of the hill they were unchained and the team pulled the front running gears ahead and dropped the load to the ground. We then put the rear-end, the hind running gears together with the front, and reloaded and chained them down and headed for home. I cut quite a few poles for other people for ten cents a pole, loaded fifty on the front running gears of a wagon and came off the side of the mountain for about a mile and then unloaded, making five dollars a day for a tough day's job. Good wages for the thirties!

Carl Iverson and I cut and snaked poles to the bottom of the big hill on the Stott place from up high in the timber in upper Ed Thompson field. These were for the Fritzes, a little north of the Tiber elevators. They had two "running gears" wagons, one tied to the other and an iron wheeled tractor with six-inch lugs for traction on the wheels. They had to go about a mile on a practically solid rock road. I'm sure it about shook the fillings out of Raymond's and Leonard's teeth. And even today with rubber over those rocks it is plenty rough. A few years earlier it would have been live horsepower pulling those wagons.

MY TRUSTY FRIEND

Arlo Graham and I were camped up Stott creek getting corral poles. We used a team and laid the butt-end of the poles on the front running gears of a wagon, chained them down with the small ends dragging on the ground. I was going around a grade above the creek and the dragging poles pulled us over the bank. The wagon and team flipped upside down to the bottom of the creek bed. I was standing on the poles driving the team, and when the wreck started, I just stepped off the load.

In order to get the team on their feet; I started cutting leather with my knife, as most of the harness buckles were too tight though some did come undone. Luckily the team was all right, but what a mess! The wagon was not damaged. With a box of leather rivets and part of a tanned cowhide to repair the harness, we were soon back in the business cutting and hauling corral poles.

Years later, Greg Weber and I went down into Brown's field to bring home Doug's registered quarter horse mare that was with Irvin's stallion. She came out from the bunch of horses up to my saddle horse. I tied my bridle rein round her neck to keep her from leaving, while at the same time untying a halter from my saddle. About that time she pulled back and broke the rein, which pulled up tight, and in a moment she was flat down and choking to death. I reached in my pocket and no jack-knife! Without a doubt this was the most "empty feeling" in my lifetime. Then I remembered I had changed Levi's that morning and forgot to transfer the knife. I was able to work my hand under the rein and get enough slack so that she got to her feet. However, she still was not getting enough air and kept falling down and getting back up. Greg caught my saddlehorse and got my fencing pliers off the saddle and cut the rein, saving the horse. It's something that should never have happened, and kind of a stupid thing for me to forget my knife.

Another time I roped a bronc in our corral and threw about three dallies around the snubbin' post to stop her. Somehow, one dallie got under another and I couldn't get it loose. The mare was down and couldn't get any air, so out comes my Trusty Friend, and I cut the rope. Yep, anyone working around horses and livestock needs to keep his pocketknife real handy.

SHORTHORNS FOR A JOHN DEERE TRACTOR

When Jim and Norma Woods bought the Lloyd Beer Ranch on Miners' Coulee, we bought Beer's cows. Sid House and I were headed to the Hick's Place with them. Dad met us at the Arthur Fey Ranch with branding irons as it was on our route, and Gordon Christian had a new squeeze cattle chute at his corrals. It was the first factory built squeeze chute I had ever seen. From that time on a lot of different makes were being built. The name of said chute was La Reau. It was mostly wood and some metal. These cows were milking shorthorn cross cows, and without a doubt they were the toughest cattle I ever helped trail. They just wouldn't trail out or even graze along. We were two days making the drive. That winter we traded them to Ernie Anderson, a John Deere Dealer in Cut Bank on a John Deere tractor with iron lugs and iron wheels.

It comes to my memory; I did get into a real "tight spot" with that tractor. I was plowing and going up a steep hill. The tractor got stuck and the motor "died" and spun out, putting the tractor back next to the plow. I was not sitting in the tractor seat, but was standing up and a long handle (the depth gage) on the plow caught my Levi's just below my knee. I was lucky it got just denim and not hide. I tried to get loose by tearing my overalls, but they wouldn't tear and I couldn't pull them out from the metal. So out came my Trusty Good Friend, Jack Knife. I was in a pretty awkward position, and didn't have too much space between my skin and that iron handle. A picture of my situation could have been a good ad for Levi Straus—"Won't rip, have to cut".

EARTHQUAKE OCTOBER 1935

The fall of 1935 we sold our band of ewes and decided we were going to run dry cattle and spay the heifers. Steers were O.K., but we had problems with what was called waterbelly. A hard white rough stone would get lodged in their water channel and the steer could not urinate causing the bladder to rupture. I have operated on a lot of steers and took out the stone, using one stitch of thread to close the channel, and several stitches on the main cut. That fall we bought Anthony Fey's heifer calves. Dad, Carl Iverson, Edgar Woods and I left early one morning and rode to Fey's and corralled the cows and calves. We had dinner with the Feys, and then cut the calves from their moms and trailed them home. It was as dark as the inside of an Angus cow on getting home with them. I remember there were sixty-three head.

The next day it was storming real bad. Dad and I were sitting in rocking chairs that afternoon (taking it easy) and my rocker started going pretty fast, and I looked at Dad and so was his. The teacups on their individual little hang-up hooks were playing a pretty tune banging against each other. It was the earthquake of October 18, 1935. It wasn't the quake that formed Quake Lake in Yellowstone; that one came about daylight, I think, in May. It shook all our chickens off the roosts and they were running round like crazy, squawking their heads off, and cackleberries were scattered all over the chicken yard. Guess it shook the eggs right out of those old hens and they had no time to go to the nest boxes.

MAD BULL TAKES A RIDE AND OTHER BULL (STORIES)

It just came to me about the time I went with Dad to get a yearling bull he had bought from Lew Berthelote who lived in the foothills next to Miner's coulee. Dad had a team and double box wagon. The floor was in petty bad condition. It wasn't good enough for grain for sure, and before we got home we found out it wouldn't hold a bull either!

The bull had a halter on and was tied into an iron ring in the floor. He got mad and on the fight and made kindling wood of the floor. We stopped in Gold butte and got some boards and slipped under him. We had to stop real often and rearrange our moveable floor, and the floor was getting slicker over the miles. A good thing the ring was secured to a brace underneath. Dad and I were sitting on the wagon seat just about a foot in front of Mr. Bull and he was blowin' snot right at our shirttails. Yep, he was pretty mad and you know what was flying out the end of the wagon all the way home. Good for veggies in the garden also makes tomatoes red and juicy. I was quite young at the time, and pretty spooked all the way home.

Years later, Claude Ritter and I were bringing a Black Angus bull that was crippled from Sheep Creek to the corrals at the Murray Place on Trail Creek. We were getting along pretty good and giving him lots of space, as Angus bulls have a pretty short fuse even when feeling no pain. However, this fellow must have been hurting 'cause he was mad at us and at the whole world. Claude was just a little kid and riding Cocoa, a Shetland pony (described by Claude as "that nasty little pony"). As we got close to the corral, he and Cocoa went around the bull to help turn him into the gate, and that little Shetland decided to try and buck Claude off. Well, that didn't work so she just laid down, and here comes Mr. Angus right for him, with Claude trying to get off the pony and up to his feet to get out of there. Luck was with us as the bull then turned away from them and went on into the corral.

Another time I was trailing an Angus bull from that area to the corrals and he was acting pretty nice, but I tried to be careful and not trust him too much. We were on a real steep side-hill and the bull was working himself too far down the hill. I went up pretty close to him, which I shouldn't have done, and he turned on my horse and ran his head under the horse catching my foot and leg, lifted us right off the ground and threw us down the hillside. To this day I can't believe that my horse didn't fall on top of me. I could have been underneath a 1200 pound horse on a real steep side-hill being cushioned by plenty of rocks. She held her footing and I was pretty lucky to escape with just a hurt leg. That's for sure, you just can't trust any of these bulls, most of all if they are black!

BUFFALO HORNS

Doug and I have found many buffalo horns over the years. While riding, your saddle horse would partially step on one and the sound wasn't like his hoof stepping on a rock. I would stop to have a look and sure enough it would be a buffalo horn. This always was in pretty heavy grass and the horn didn't show up unless it was disturbed. I have a pretty good collection. The most interesting one I have was found by my Dad in 1903 on Sage Creek. In the horn is embedded an old slug. The buffalo must have carried it around quite a few years, as the horn had partially grown over the slug. Dad gave it to Uncle

Claude who took it back to Nebraska with him. One day Uncle Claude caught his kids starting to paint red around the slug (to look like blood?). He decided that the horn better return to Montana where it would be more appreciated, and returned it to Dad. It still has the red paint on it.

A FEW STRAY THOUGHTS

I have always said "A man's home is his castle, whether it be a dug-out in the hillside, a tent, a sheep herder's wagon, a tar-paper shack or just a modest roof over his head".

On a ranch in the early days, if a man could harness a team and hitch them up and go to work, or could saddle a horse and step on and go to work, that is about all the ranch education he needed to get by. Of course if he married, having a helping partner and a good cook was a big bonus too! Now days you need to know that and more, such as a little know-how on being an electrician, mechanic, plumber and carpenter as well as having a fair education. You need to keep up on new methods in agriculture and of raising beef. You have to have a good "business head". You also need a real sharp pencil, 'cause most Bureau-

Demarest Ranch

crats are not worried where their grub comes from and slow at getting things done. If you told them that sex caused cancer, it would take them forever to figure out where to put up the warning sign.

My favorite color is green.
Green Grass,
Green frog-skins,
Green broncs (earlier in my life),
And Mountain Dew!

Demarest Ranch

Three Generations, Harry, Claudie and Doug

VII / Fifty Years of Spaying

MONTANA SPAYING EXPERIENCES

In 1936 we decided to run spayed heifers. At that time they were going to the Corn Patch as two and three years olds. We called all over the State of Montana to find a vet who would give me some pointers. There were lots of vets, but none knew anything about spaying, or maybe they didn't want any competition. In the early 1900's into the 20's it was done quite a lot. An old time spayer, Everett White, from Augusta, Montana came up to the ranch and got me started. "Keep in mind", he says, "you can't be too clean, after all it is an operation". So the next few years I practiced on our own heifers. The first ones I spayed were 63 head that we had bought in 1935 from Tony Fey. My Granddad John Brown made me a long handled tool, hook shaped on the end. I am still using that same tool today and the hook is worn pretty thin after over 50 years of spaying. Now they do have a spaying scissors. I have a pair, but always liked my homemade one the best.

I soon found out that if a twin heifer's twin was a steer, she didn't have ovaries. She is called a Freemartin. If twins were both heifers they had ovaries.

The Flying U Ranch at Choteau called me up and asked me to spay their heifers. I refused them saying that I still didn't feel I should work on other rancher's cattle. They wouldn't take "no" for an answer, so they were my first outside cattle to work.

As these heifers are put in a chute for the spaying procedure, I just want to explain a little about chutes here. The Sheridan Ironworks from

Sheridan Wyoming made the first patented chute called "The Johnson Senior Squeeze Chute". I ran into quite a few of them in my travels. The head catch, which wasn't a stanchion type, wasn't too good. The squeeze worked with two big springs on each corner, and the whole side narrowed up top and bottom working together with a rope on a two-by-four overhead that operated the squeeze, and a little lever just above your head, that worked as a friction on a rod to release little by little as you touched it. The tailgate pulled or pushed in or out on a track. The hoof trimming and milking panel at the bottom let down with hinges with a real easy working latch. Except for the head gate it is the most perfect chute made to operate. We still have one at the ranch. Quite a lot of this chute was wood instead of metal which made it much less noisy. After 70 years of use, it has seen its better days. There were very few of these patented squeeze chutes for some years. Most of the ranches had old wooden ones. I am sure I have worked with every brand of chute made. If you took the good points from every brand of chute and built them into one chute it would be the perfect one.

For quite some time I used Lysol for disinfectant in my three different containers of hot water and tar put on with a paintbrush on the stitched up wound. And now sulfa powder is put in the wound and scarlet oil stops all drainage. Now there are better disinfectants to put in the water. There is a powder called "Poly Powder" made to put into drinking water for baby chicks. About 2/3 of sulfa powder and 1/3 of this has a faster healing effect. I used Button and Carpet thread to sew them up. I still use tar, as it is great to seal wounds. It was a great help when I could use electric clippers. When I started many of the ranches did not have electricity, and I had to use the hand clippers. Also many cow camps and moveable corrals would be miles and miles from power.

People ask how long does it take to spay a heifer and about how many did you work every spring? After a couple of years getting "the know-how" accomplished I would average twenty heifers an hour or three minutes average per animal. That would be an all day average from morning to night. That time included doing all my own clipping and then washing the area with a solution of disinfectant and hot water. A person at the chute control put the tar on because it was kind of messy and sticky. Every spring I spayed between 7000 and 8000 heifers.

Many ranchers who saw our spayed heifers in South Saint Paul Commission Company were really impressed at what they saw. They were smooth, fat, beautiful heifers. So seems like the next spring I'd be at their ranch, meeting new people and seeing new country. I drove miles and miles and partly at night too, from place to place from the Rocky Mountain Front south to Wolf Creek and east to Glasgow and all along the Canadian border. I also did a couple of bunches in Canada. Oh, oh! Hush, hush!!

My biggest spaying jobs were at the Hamilton Ranch at Simms, Montana on the Sun River. They ran just dry cattle and had heifers to spay both in the spring and in the fall, with an average of 1000 to 2000 a year. I spayed mostly at the home ranch and also at The Red Shed, and the New Barn at different times of the year, and sometimes at the Gibbons place located on the city limits of Choteau. The Hamilton Ranch is now owned by the Broken O. I wish I'd kept all my stubs from the checks that were written to me during 25 or 30 years of spaying for the Mangus Cattle Company (The Hamilton Ranch of Simms, Montana).

When working heifers, if one looked sick or didn't look healthy, I wouldn't spay her. One day this particular heifer went out not far from the chute and dropped dead just as my boss, John Hamilton, walked around the corner of the holding lot. I was a bit embarrassed, but he really had a laugh. So, I guess that was one that wasn't too healthy and the shock did it. Anyhow, she was never going to see the feedlot in Bakersfield, California. That's where all the Hamilton cattle ended up. One spring the fellow that owned the feedyard came out to visit the Hamiltons and wanted to see how it was all done to produce such beautiful fat cattle. Another time at Hamiltons a heifer got part way out the chute. Irvin Brown was there and he grabbed her and bulldogged her to the ground and with some more help from the chute got on her, so I just spayed her lying down. And believe me it was a bit awkward. John Hamilton came by during this episode. He laughed and wondered if the chute was just to show or if it was wore out and maybe he better get us a new one!

In the 50 years of spaying outside cattle I was very lucky and never got bruised up too bad. I had a lot of respect for those working handles and levers on these patented chutes. They did the job, but if you got smacked by one it did not bend that easy. I had worked about 700 heifers at the Gibbons place on the edge of Choteau and came down to the home ranch at Hamiltons and got started on about 500 head. I had only put about four or five through and I cut my hand real bad in the fleshy part between my thumb and finger. It was pretty deep. So I went up to the cookhouse to get Elsie Norris (she was the cook and Ray Norris's wife; he was the foreman on the ranches) to bandage me up. Right at that time Mrs. Hamilton came in and she got pretty shook up and says, "I'm taking you to Choteau to the doctor and to get a tetanus shot." So we did. When we came back I looked up Ray who was up the river working cattle. He said, "Don't worry about it. Go home and heal up and come back." It just so happened it was into June and these were some heifers they had shipped up by rail from their ranch at Silver City, New Mexico, and it was the last job I was on that spring. We always did these heifers from that ranch in June. They were real nice yellow haired Hereford cattle.

One fall when I arrived at the Hamilton Ranch at Simms they were not quite ready for me. I was glad because I got to help them gather 6000 yearling steers and some heifers to be spayed. We just worked them into a big holding pasture and dribbled them into a working corral where they took out 1500 heifers to be spayed. They gave me a mouse colored gelding to ride with a "Fish Tail" brand on his thigh. They had quite a few saddle horses on that ranch with that brand bought from a ranch in southeast Montana. I had ridden him before. I always carried my saddle with me for helping out on different outfits if they were short handed. I have always cherished that experience to have had the opportunity to have helped work that big a string of cattle on grass away out in the open country.

One spring I was at the Hamilton home ranch spaying. Ernest Price (he was an old friend of mine also raised in the Sweet Grass Hills) was working there and he and I saddled up and after supper rode down

Sun River on the north side. He had a gate he wanted to shut. I was riding the little mouse-colored horse and he was riding the tall mouse-colored horse. They looked just alike except for size. We came to a road crossing on the river. Ernie said that it was a good solid bottom crossing, and there was wagon tracks where someone had crossed maybe the day before. Anyhow, ole Sun River was pretty wild that evening and in flood stage. Says Ernie, "Let's cross here and go back to the ranch on the other side," and into the river he and the bigger mouse colored horse went. And I knew for sure I wasn't going to cross the river, 'cause belly deep is too deep for me. Anyhow, about half way across his horse hit a terrible current, and all that was showing was his nose. Ernest had quit him and was hanging onto the horn. The horse was smart and frantically turned upstream and was fighting the current. Twice Ernie yelled, "I'm a goner, I'm a goner". Finally the horse fought himself out of the swift current and they made it to shore. He shouted back to me, "Don't try it, don't try it!" He didn't know I'd made my mind up before his horse was knee deep which side of the river I was staying on. It was the most terrible thing to watch and be so helpless. Like he said, that if that horse hadn't kept his senses and fought against the current and had drifted downstream, they would have drowned. And he darn near did anyhow. When they got to solid sand, Ernest laid there quite a spell. The horse also was all in. Ernest's legs were pretty badly bruised up from the horse's feet hitting him as he'd had to be careful not to tip the horse on his side from his having hold of the saddle horn. Several hours later we both got to the home ranch. It got dark and I kinda got a little lost going back. Anyhow, I was dry. I didn't get to see much more of Ernie because he most of the time was camped at the Burke place.

Aron, the chore boy, came tearing out to the chute one morning trying to locate a milk cow's heifer. He said that the boss didn't want her spayed, she was to be a milk cow some day. I told him it was too late, she already went through the chute. He was pretty happy and said he didn't want her for a milk cow anyhow!

About ten every morning a big white Leghorn hen would show up and start picking up ovaries from the ground around me. She could hold a lot of them. After about half a day of feasting she would cackle her way back to the flock and that pretty high cone of hers to me seemed a lot brighter red when she left, and I'm sure she laid a lot more cackle berries than the other hens!

As soon as Patsy Hamilton was home from school she always came out to the chute and kept my needle threaded, 'till we heard the "first" bell ring for supper.

At Hamilton's is where I got acquainted with Bob Sauke. He was cowboyin' there and we had some good visits, a great fellow to know. Jerry Buckley was also cowboyin' at the ranch one spring when I was there. I remember he was getting on his horse inside the barn that had a plank floor. The horse stepped on his foot, and his other foot was already in the stirrup. Jerry was kind of upset, and I got to laughin' so much I had to leave. I don't think Jerry saw the same humor in his predicament as I did.

The chore boy at Hamilton's had a couple of Labrador dogs and on one of my trips she had a nest full of puppies. He kept her in a stall in the barn. One evening he said, "Come and I'll show you how smart she is." In the batch of puppies was a little "runt" and not able to find his "plate", as he just wasn't that healthy and active. He let her into this closed box stall, and she nosed through the batch (I think there were ten) and got the little runt by the skin of his neck and carried him to the far corner and let him nurse till he was filled. She then took him back to the nest and then she let the rest of the puppies get their supper. He told me I could pick any female puppy, so I did and brought her home to Auverne and Doug in a shoebox, she just fit. We named her Cindy and we had her for a long time.

I went to Roland Peters at Bynum and their heifers were in a lot and being fed dry chopped hay and there was no way I could get to the ovaries or take them out. They were just too tight inside. So we had to take them off feed for 24 hours. Green grass or hay would have soon gone through them and would not have packed them so tight for very long.

Another ranch where I worked a lot of heifers was at the _F west of Augusta. Tott Nett was the foreman at that time. They also had a cow camp south of Great Falls around Sand Coulee and most springs the heifers were down there as they were summered down in Tiger Butte area. One evening Tott had Pete Jarvis and Shorty Henry go into Augusta for some repair and I went along. Tott's little boy, Terry, (about four years old I'd guess) was also with us. So before heading back to the ranch we went into the Buck Horn Bar to have "a little toddy for our body". Little Terry sat on a stool next to me. Shorty (Terry's uncle) tells the bartender that Terry is going to buy the drinks for us. And the little guy just kept sitting there not saying a word, just really ignoring all the goings on about him buying. Pretty soon I noticed him working his little fist down in his pocket and he came out with a hand full of pennies and he slaps them down on the bar top and says, "Div me a toke, to hell with them three sons-a -pitches". I was telling Tott the next morning how little Terry pretty much took care of the situation. "Yeah", Tott says, "he spends a little too much of his time around the bunkhouse". Years later Terry went to vet school and I understand he is way up in veterinarian science and experiments.

John Robinson south of Chinook on the south side of the Bear Paws was also one of my customers. I knew John before I knew his brothers, Jess, Don and Jim, and always looked forward to the drive east into his ranch to spay. John and Herb Williamson were neighbors. Herb was an uncle to Larry Kane, a professional bronc rider a few years back.

The 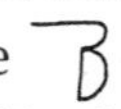 Cattle Company south of Zortman in the Little Rockies owned by brothers Jess, Don and Jim Robinson supplied me with a lot of material keeping me plenty busy for a spell. One day at dinner we had a big kettle of mulligan sitting in the middle of the table. Nancy Robinson was in Malta that day so dinner was bachelor style. Anyhow, this Indian boy, a long time hand on this ranch with plate in his hand and a big spoon says, "Dig deep, puppy in the bottom!"

Jess and Nancy's son, Clyde, and Don and Eva's daughter, Diane, learned to spay during different times I was there. Bob Fry who ranches south of Malta does his own heifers now, also. I also taught Glenn Thompson to spay heifers when he worked for us. Robert Dangerfield from across the Border spayed a few of his own also when I was at Art Wardell's ranch.

When spaying at Jess Robinson's ranch he always kept his rope horse tied up close by the cattle chute. Jess worked the head gate and quite often a heifer would escape and Jess would step on his horse and go rope her and bring her back to the catch pen. And you know Jess, I have always figured it was kind of planned. A good roper and his top horse have just got to stay in practice! Right Jess? What a better time to keep from getting bored at the head catch!

Up next to the Canadian border north of Hinsdale lived Harley Hunter. I had met him the fall before in Saint Paul and he told me to come the next spring and work his and his neighbor's heifers. Elsie was with me on this trip. The previous spring his neighbor, Ophus had a vet spay and said that they didn't have too good of luck. That was all he said and I didn't ask any particulars. While I spayed the first two or three heifers the crew was crowded around me so close to get a good look, that I hardly had any elbow room to work. Then they told me that the vet the spring before had them hang the heifers by their hind feet upside down, powered by a farm hand and a tripod. Well, I guess it was a terrible mess. I don't remember now all the details as for numbers and days it took, but they didn't all live to make it to the Corn Patch back east.

Mr. Ophus was interested in learning how to spay and with my direction he spayed a few heifers. He took a copy of my spaying hook and later had one made for me. I was pretty happy with the gift, 'cause by that time my spaying hook was getting pretty thin mostly from being honed to keep it sharp, and from removing lots and lots of ovaries. While driving from one job to the next, I happened to stop at the sales ring in Shelby and being not too rushed figured I'd relax for the afternoon and watch the sale. A bunch of yearling heifers came into the ring and a buyer remarked to Gibby Gilbert, the auctioneer, "if those were steers I'd give you a bid." Gibby points at me and says, "There is a fellow in that top row up there who can take care of your problem. Go talk to him 'cause we got a lot of heifers to sell but no steers" So out at the stockyard squeeze chute I was kept busy the rest of the day and into the next.

I also remember working some grass green heifers in a feed lot at

Vaughn, Montana for Dave Dear, and at another feedlot on grass green heifers in Conrad for Willard Nauman.

I was in Augusta and called a rancher down by Stanford. He said his heifers were pretty fat as they were fed a little grain during the winter. So I told him to corral them the night before and they would be easier to work. Otherwise I didn't want them too empty 'cause a cow critter is a lot easier to get along with if she is not too hungry. Guess that's the same way with cowboys, too! I pulled into his ranch in the evening and stopped and looked at his chute and corrals, and was pretty happy when I saw a bright red squeeze, a "Powder River", and the control levers were on the right side. There was a big pile of cut wood and a big stove by the chute. Then I drove on up to his house and he asked me how much help we needed for tomorrow and how long it would take. Well, I told him that with a hundred heifers we would be through by noon and two helpers, one to push them up the alley to the chute and one to operate the controls were all the help I needed. "Oh, my God!" he exclaimed. "I got half the neighbors coming to help and I figured it would take about a week. And I knew you used hot water and that is the reason for all the wood and stove". What I mean, he was really prepared!

After dinner I left that ranch and drove down on the Missouri River about where the town of Craig is and was met there where the river is close to the highway. I took my suitcase with my tools and crossed the river on a raft with the owner of the heifers who was the navigator. The heifers belonged to Charlie Lincoln who had some rented pasture in that area. He did not live there. His place was on the Marias River. Now there is a bridge across the Missouri where I went across on not too good a raft. Where the corrals and cattle chute used to be is a big bar and nightclub and "nose bag".

Another time at the Charlie Brinkman ranch on the Marias River, which was being operated by Les Taylor, I had to cross the river where there was no bridge. Les led an extra horse for me and my suitcase of tools and crossed me from the north side of the Marias River to the south side. It wasn't swimming water, but I did get my boots full. I was kind of wishing I'd driven the extra forty miles around to the bridge, 'cause every three weeks on a Saturday night was my time to take a bath, and I still had two weeks to go, besides, I can't swim.

I remember a string of Scotch Highland heifers that I worked for Bill Cowan at Box Elder. Their hair was a good six inches long and after all that clipping it left a pile about enough to fill a hayrack.

At one ranch the cowboys back in the catch pen decided to play a joke on me and ran a steer into the chute for me to spay. And I guess they were a grinnin'. When I put my hand inside, I say to myself, "This is either a steer or a Freemartin." So I check under the critter's tail and it was a "one-holer", not a "two-holer". But it did happen once in awhile that by mistake a steer would be in the bunch and not noticed.

I spayed some heifers for Kolstad south and east of Chester and while at work the county agent (I think he was the agent) from Chester came out to have a look. Yep, he took one look and fainted and fell over and in doing so he hit his head on the hind bumper of my car. He was out cold for a spell. He said afterward that he knew he really shouldn't even have showed up to look. Anyhow, the first that I knew he was even there was when he hit the ground. He left with a black eye and a bump on his head, after I washed up and had a Chesterfield and a little visit. On leaving he says, "I'm not even going to show up at the Court House today."

Donny McGillis was telling my son, Doug, a few years ago at the sales ring in Great Falls, "The first time I met your Dad, he was working heifers at the home ranch at Hamiltons and I was shoeing a horse by the blacksmith shop pretty close to the cattle chute. I had a little fracas with my horse and he kicked me in the butt. I dropped my Levi's to see how bad I was wounded and your Dad run over and throwed a bunch of sulfa powder on my cut and says, ' Just pull up those Levi's and go to work. This good stuff will take care of any tough cowboy', and you know it really worked. Anyhow that's how we met. I guess I should have felt lucky that he didn't want to take a couple of stitches with that Button & Carpet thread!"

After I started spaying a lot of vets started as it was getting popular. Ranchers could run spayed heifers with their steers and at that time they were all kept until twos and threes before shipping to the Corn

Patch. This went on for many years until finally the housewife didn't want that big a cut of steaks after being finished. Now our 600 pound calves go to be finished on corn and of course at that time our calves were around 350 to 400 pounds off the cow at shipping time.

One spring I get a letter from The Board of Veterinary Department of Professional Licensing. At the Veterinary Convention this one vet says, "There is a cowboy with no license who is spaying thousands and thousands of heifers in Phillips County and he is also teaching others. He has got to be stopped." So this letter states I was subject to a fine plus a little time in jail. Our attorney said being I had never "Hung Out My Shingle" as a vet, I'd get by just being a hired hand for the rancher where I was spaying. Bill Hersch, our vet in Havre, was at this convention and of course knew what cowboy they were talking about. In fact, Bill got me started in that area of the state, taking over some of his customers as he didn't like to do it that well and didn't have the time as his area took in a lot of country. Anyhow, for several springs I'd get a reminder letter from that office in Helena.

South of Landusky at a four corners, I drove into a filling station with a connecting "Water Trough" planning to get gas and a little toddy for my tired body. This was south of the Little Rockies. A guy filled me up and I started to go into the place to pay him, and he stopped me and says, "Don't go in there. A pretty wild bunch are in there and really getting out of hand and I own the place and I'm not going back in either." I told him, "Well, I don't have any cash on me, I'll have to give you a check and you don't even know me." This owner replied, "Oh yes, I remember you. You've have been here before. I'm going over right now to a neighbor and call the law and have them take over, so I can go home". And you know, a couple years later I stopped there, got out and stretched a little and wondered how tough this place was this time. I looked around and it was burned to the ground. No gas or toddy for the body this trip through.

I drove into the McConkey Ranch in the Bear Paws, south and west of Big Sandy. I had spayed there before and knew them well. Started to put another nail in my coffin and Leon says, "Here try one of these filter tip cigarettes, they are just new on the market, but I know you are not going to like it. They are just like sucking on a tittie through a flannel nightgown." "Leon, at your age you should know." So I lit one up and then said, "Yes, I agree with you, you're absolutely correct!" Robert McConkey and his brother, Royal, were Leon's sons. Royal was on the ranch with his Dad and Robert had a different ranch close by. We always used his chute when we worked their heifers. At that time a little kid by the name of Warren was a little too young to be of much help. Today Warren McConkey is manager of the Flathead Electric Co-op at Kalispell. Makes me feel kind of ancient.

A rancher in Washington State wrote me a letter asking if I could come and spay in his area, and that he could furnish me with all the work I could handle. I also had a chance to go into northern Wyoming one spring. Traveling to those other states would have been great opportunities to see new areas. However my home state customers kept me so busy there was not time for out of state travels.

There is a cow camp several miles west of Belt, called "The Sand Coulee Camp". It was owned by the J. B. Long Ranch at the Rocky Mountain Front west of Augusta and operated by Tott Nett. Some springs the heifers were at one place or the other at spaying time. There were lots of rattlesnakes all around you, really had to watch your movements. One morning one of their saddle horses had got bitten on the nose, which swelled up as big as your hat. Pete Jarvis, an Indian cowboy, was camped there and one forenoon while we were working he kept talking about the big beef roast he had a cooking for our dinner, so a little before noon he left us and went to the log cabin to fix our chow. When we walked in he was opening some cans of Spam for dinner. No beautiful roast. It was burnt to a crisp about the size of a silver dollar and just as hard. Once while coming down a long steep road into Sand Coulee cow camp one of the cowboys from there was riding with me. Off to one side in the ditch was an old wrecked car. He looked over at this outfit, and real serious says, "Yep, nine more payments and she will be mine", real proud like.

On one of my spaying trips I stopped over in Great Falls and stayed overnight at the Johnson Hotel. The next morning up on the third floor waiting for the elevator, I met another rancher who I knew from

Choteau. I had recently spayed his heifers. The elevator stopped and we crowded in. When we started down he noticed that the zipper fly on his pants wasn't zipped up. Right close in front of him was a lady wearing a fox stole and it was hanging real low from her shoulders. Some hairs on the tail of the fur piece got caught in his fly when he tried to zip up, and for three stories down he tried real "easy" to get loose. When we landed at the bottom floor he kind of stood his ground thinking that when she left a few of the tail hair on the fox hide would break loose and he would be free. Instead the foxtail came off the hide and there he stood in the open door of the elevator with the foxtail hanging between his legs. The elevator faced the hotel lobby with a pretty big crowd of spectators. The gal whipped around to see what just about upset her. My friend stayed pretty cool and collected and just jerked the foxtail loose from himself and handed it to her. After watching that little happening (with a ringside seat) the thought or sight of a zipper brings a big grin and a chuckle, but kind of scares me also. But the old saying is, "If ifs and buts were fruits and nuts, oh, wouldn't we have a Merry Christmas!"

I was spaying on a ranch down next to Tiger Butte. After getting started the owner noticed our turned-out heifers were crossing a snow-bank into a field that he didn't want them in. He went after them and a terrible electric storm suddenly came up. We could see him as he was up on higher ground. We saw his "white" horse hit the ground from a bolt of lightening and in seconds he was on his feet again. Our Boss was on his feet too, and gets on and rides back to the chute and cattle. He was a little pale. I also got a bad electric shock during this storm. I had my hand inside a heifer and it felt like it curled up, and a feeling that my toes curled up a little, too. I sure backed out of there in a hurry. Anyhow, my help and myself waited till the storm was over. I'm sure that having an electric storm when there are still snow-banks doesn't happen too often.

The fifty years I did outside spaying I was averaging 7000 to 8000 head in spring season besides a few in the fall. It was a great experience meeting a lot of ranchers and traveling all over Montana. It was also good wages at $1.00 a head for this cowboy who had punched many herds of cattle for $2.00 per day. The Hamilton Ranch paid me 60 to 80 cents per head depending on the amount in the bunch, and they furnished all the disinfectant. My motto was, "I treat every heifer like I own her".

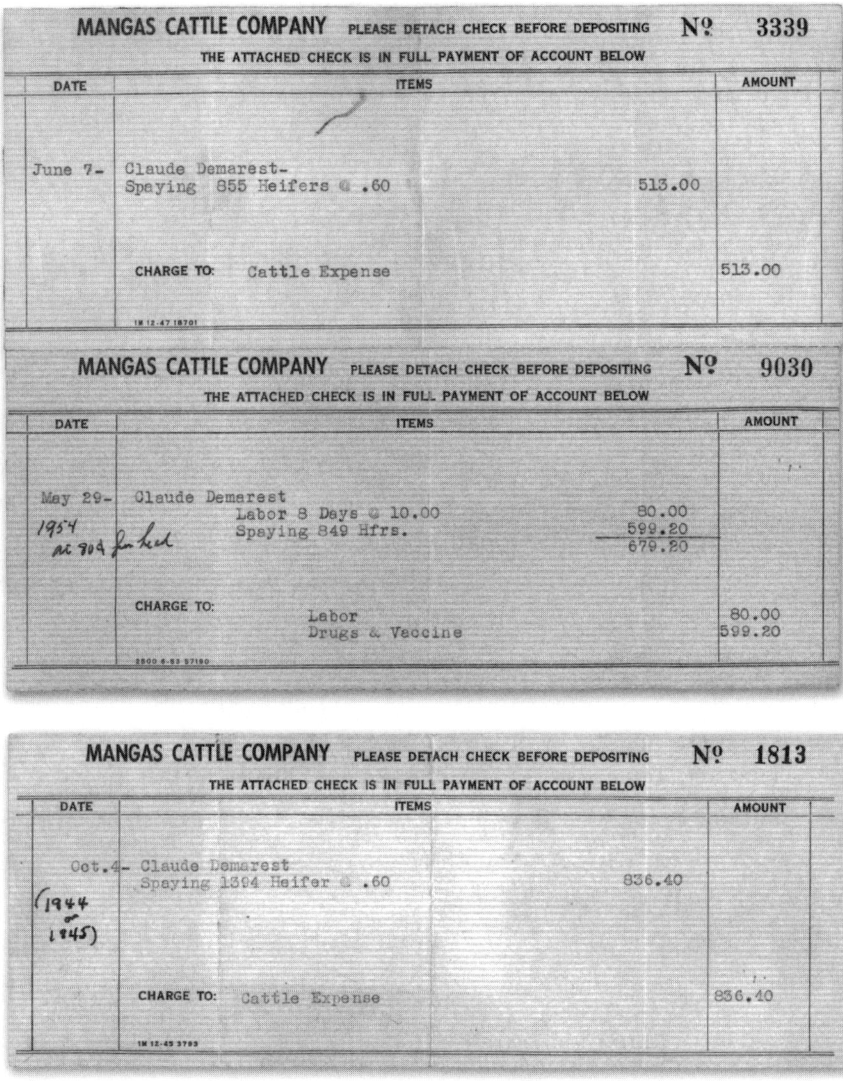

MANGAS CATTLE COMPANY PLEASE DETACH CHECK BEFORE DEPOSITING Nº 3339

THE ATTACHED CHECK IS IN FULL PAYMENT OF ACCOUNT BELOW

DATE	ITEMS		AMOUNT
June 7-	Claude Demarest- Spaying 855 Heifers @ .60	513.00	
	CHARGE TO: Cattle Expense		513.00

MANGAS CATTLE COMPANY PLEASE DETACH CHECK BEFORE DEPOSITING Nº 9030

THE ATTACHED CHECK IS IN FULL PAYMENT OF ACCOUNT BELOW

DATE	ITEMS		AMOUNT
May 29- 1954 at 70¢ for head	Claude Demarest Labor 8 Days @ 10.00 Spaying 849 Hfrs.	 80.00 599.20 679.20	
	CHARGE TO: Labor Drugs & Vaccine		80.00 599.20

MANGAS CATTLE COMPANY PLEASE DETACH CHECK BEFORE DEPOSITING Nº 1813

THE ATTACHED CHECK IS IN FULL PAYMENT OF ACCOUNT BELOW

DATE	ITEMS		AMOUNT
Oct.4- (1944 or 1945)	Claude Demarest Spaying 1394 Heifer @ .60	836.40	
	CHARGE TO: Cattle Expense		836.40

Check stubs from Mangus Cattle Company

VIII / Community Life

EARLY TELEPHONE SYSTEMS

The year Mother and Dad were married was 1911, and she kept a diary. In it she mentioned on June 28th, 1912, the phone line came into their home and was installed. The phone was bolted to the wall and had two dry cell batteries inside. On the side of the phone box was a handle or crank you turned to put in your call. Chester (central) was one long turn on the handle. Ours was one short turn of the handle and one long turn of the handle or crank. It was a two-wire line and started at Chester, went north to the Laird Ranch, then to Whitlash and on to Gold Butte, which was the end of the phone line. There were lots of lines branching off to different ranchers. It was called "The Sweet Grass Hills Telephone Company".

In later years our Ranch, the two Brown ranches and the Stott Family at Whitlash had a telephone system that ran on the top wire of a barb wire ordinary fence. At wire gates it was carried overhead with two tall poles. Wherever there was a splice previously made to connect the barb wire, they used a short piece of new galvanized wire to go around the rusty splice on the main wire. At that time the two Brown ranches never hooked up to the main telephone system. That is why they had the barb wire line to our ranch and to the Stott family. So the Demarest Ranch and the Stott's had two telephones on the wall of the house, one for each system. That barb wire telephone line worked very well. Yet today on our ranch, some of those galvanized splices are still there on the top wire, which now only turns cows, and no longer carries voices.

Telephone stock certificate dated October 21, 1912

NATURAL GAS AND ELECTRICITY

In 1927, '28, and '29 natural gas wells were drilled in the Sweet Grass Hills. The West Butte coal mine ceased to operate, and I understand there was plenty of coal left down under for the taking. Our houses then got piped in gas and we began to heat and cook with gas. We also had gas lamps that were attached to the walls. Later we used Coleman gas lanterns, which we hung from a hook in the center of the room. They were much brighter than the gas wall lamps.

In 1929 Montana Dakota Utilities piped natural gas from the Whitlash Gas Field to Great Falls. Up on the hill west of our ranch they built a "booster plant" consisting of big motors to help push the gas down the pipeline. This was before four-wheel drive pickups and our ranch rented saddle horses for the two men from there who made rounds of the gas wells to pick up daily charts. Harry Gardener Jr. was one of those employed there and rode one of these horses. Harry and Velma (Elsie's sister) lived at the Gas Camp close to our ranch. I kept those horses shod and grain fed and made them "Step Out Head Up and Tail a Risin'".

Next we bought a Kohlar light plant that operated on natural gas. This would have been in the late 1940's. We had the houses wired for electricity and when you pushed any light switch in the house, that would start the Kohlar light plant automatically. We had this light plant for a few years before we got electricity from the Rural Electric Association, which came in 1951. Except for keepsakes, out went the Coleman gas lanterns and the coal-oil lamps.

I was a Trustee of the Marias River Electric Co-op and attended monthly meetings in Shelby for many years. In March, 2001 I "bowed out" and resigned after 48 years as one of the Trustees. It was an educational experience. Mr. McPhillips, Bill Current, and Larry Nelson were different General Managers during that time. Larry is at present still General Manager. Since 1996 Marias River Electric has had the cheapest rate per K.W. H. in our State of Montana Electric Co-ops. At my age some "new and younger blood" was needed, and I guess I was getting like an old saddle horse that didn't want to leave the barn. At our annual meeting in April 2001, I was presented with a beautiful plaque congratulating me for my 48 years as a trustee. Adrian Hawks was voted by the members to fill my empty seat.

MAIL ORDER CATALOGS AND A TRAVELING SALESMAN

When I was a kid growing up, if it hadn't been for Sears Roebuck, Montgomery Ward, National Bells Hess, Spiegel and other catalogs, we would not have been able to do our shopping very easily. These catalogs kept everyone's family dressed for fall, winter, and spring. The distances you had to travel to Chester (40 miles), Shelby (60 miles), Havre (100 miles), or Great Falls (150 miles) prohibited shopping on any regular basis. At first, people only had horses and wagons. Then the early cars didn't travel well on our dirt roads. When rained it got very muddy and the gumbo rolled up on the tires. In the winter the snow got plenty deep on those roads, which were not elevated as they are today. There were no such vehicles as four wheel drives, pickups or four wheelers. It was quite a few years until we had those kinds of vehicles and the improved graveled and oiled roads that we have today.

Our food was bought in town in large quantities; 100 pound sacks of flour, and sugar in 25 pound sacks. I remember dried fruit such as prunes; apples, apricots and peaches came in 25-pound cloth sacks. Much of our dried fruit came from Saxtons, a mail order house in California. The cooked dried fruit was more tasty and flavorful than the canned fruits we buy from our stores today.

When these mail order catalogs were a year old they sent us new ones, besides we received the summer, and winter seasonals. I must add that when the catalogs were out of date, they still had a use in The Little House out back. Softie pages were used first!

Torger Forseth, a brother of our Aunt Randi Brown worked for The Minnesota Woolen Mills. He would travel from ranch to ranch showing samples of clothing and taking orders. About a week later in the mail came your comfy warm duds. He got us natives fixed up for cold weather and this kept us cowboys from having an excuse to get into town where we might "forget" about buying our warmies! He also sold wool blankets, and I still have the lavender one that Elsie purchased from him so many years ago.

PRAIRIE FIRES

I have helped fight so many, many grass fires. In days gone by, all we had was a wet sack or anything that was handy in a hurry, like a piece of coat, or even your long johns, and you hoped someone would show up with some water to wet the sack. Most of the fires were caused by lightning strikes, and a few were garbage barrel fires. One was at the Wallace Ranch and it went to the top of the Bingham Ridge. The community finally got a big chain drag about six or eight feet square. It was a cluster of entanglement chains woven together. Ropes were attached on two corners of the drag and it was pulled with saddle horses. Up the line of fire you would go, one horse in the burn and the other on the grass.

In the fall of 1929 a big fire started a little east of the Brower Ranch on the big lease in Canada. We fought that fire till it went into Milk River where the river crossed the border into Montana. It lasted several days and nights. People followed along with old cars and teams and wagons feeding the fighters and giving them rides. Lots of saddle horses had to be changed off pretty often. I know without a doubt that was the most "all in and done out" everyone had ever been. Most of the help came from across the line in Montana because the fire was pretty close to the border. I was riding a horse a good share of the time on one corner of the drag and on foot at times following the drag. By the time it was over I had ridden many different people's horses. The fire drag put out the main flames, but someone still had to follow-up. After that big fire with smoke and heat and no sleep for two days and nights, most everyone's eyes looked like a couple of calf nuts floating around in a bag of blood.

A brown beer bottle started a grass fire one day in Canada just across from the Aden Port of Entry. Another fire on the ridge southeast of the Joe Roke Ranch in Precott's field was also started from a brown beer bottle. So I guess it shows that enough sun heat with the right conditions can burn a lot of grass.

One spring in April we had a bad prairie fire that started in the north end of the Simmons Place. It was started by a crew of welders who were working on a gas well. The wind took it south to the Simmons coulee, where it stopped as the coulee was running water from the spring snow melt. I got to the fire with my 1951 Plymouth car, and to be safe, I parked it out on the burnt part of the grass. I jumped out and started beating the flames. I was in too much of a hurry, and a few minutes later heard a big bang! Careless me, I left one hind wheel parked on a burning cowpie. I had to use some water that was in the car to put out the burning cow pie and what was left of the tire. Lost some air and the tire, but luckily it didn't do more harm. The old one room log Simmon's house was in the fire's path and that old homestead went up in smoke and burned to the ground. All that was left for the old homesteader who had cut and peeled the logs, sawed them into the right size lengths and put them together were memories.

Another time a drilling rig in the northwest corner of the Bill Townsend Ranch caught fire and burned to the ground. Rigs at this time were all wood construction. The fire started a grass fire with a wind in the west and it was headed for their home and buildings. People went and got Mrs. Mabel Townsend out of her house and to safety, as she was a bedridden patient and seriously ill. We were lucky and got the fire out a few hundred yards from their buildings. I have even fought fires in winter when there were snowbanks.

In the days when the whole community was still on a single party line, if there were a fire, you would ring four long rings on the phone. Everyone who heard this ring would get on the phone to hear where the fire was. My kids can remember this. Usually the men were away from the house, so Elsie, my Mother, and the kids would gather up all the cream cans and fill them with water, grab the gunny sacks from the granary, load it all in the car, and head off to the fire. Someone also had to alert those of us who were out in the hayfields to come and help. The women and older kids also fought fires by beating at the flames with the wet gunnysacks. In other words, "all hands and the cooks". It's still that way yet today.

Now days we ranchers and farmers have tanks of water and a motor producing lots of pressure to spray the fire. Before, there weren't any pickups or water tanks. Most of the water was hauled in barrels in wagons, cream cans in the car, or whatever worked. Water was used to keep your sacks wet and for a good drink for the body if nothing better was handy. These days on our ranch we keep fifteen to twenty gallons of water in containers on all the pickups. There are five-gallon containers of water on the haywagon, baler and swather. Just an "Ace in the Hole" in case of fire.

ICE HARVEST

Before we had electricity for making ice in a refrigerator, we had an ice harvest every winter. When the ice in a lake got about 18 inches or better thick, we marked out squares on the ice using a wooden 2x4 as a straight edge, and using the point of the ice tongs to cut a line about one inch deep. The squares were about 24 inches or whatever sizes you wanted the block of ice. We used a special saw made for ice cutting. It was six or eight feet long and had two handles about ten inches apart at one end. The double handles were made for both hands of one person. Then we pulled the block of ice out of the water with ice tongs, put a rope on it and pulled it up a little chute into a sleigh or truck. Our icehouse was a big hole in the ground with a low shingled roof. We slid the blocks down the chute into the hole and they were fit close to each other. Straw was piled between the layers. When we had the hole full of ice we covered it all over with more straw. It kept real well and we had ice all summer. Most kitchens had an icebox. A block of ice was put into the bottom compartment, and there was a drip container below the ice to catch the melting water. The food was kept in the compartment above. Everybody had hand crank ice cream freezers and we made our own ice cream. So, there you are, "up town" ice cream using ice, and ingredients harvested and produced in the Sweet Grass Hills. Our Jersey milk cow who gave real rich cream had a big part in the ice cream making. It all took the place of the Schwan Man's stop of today.

As money was scarce during those times, I also helped the Strode and Johnson Ranches put up ice for several winters in order to make a few extra dollars.

My daughter, Auverne who can remember the old icehouse and making homemade ice cream, writes:

"These days Dad buys his ice cream from the Schwan Man who drives out into the country and makes rounds of all the ranches. It's excellent ice cream, but it will never replace the 'home-made' variety, which was really appreciated and a wonderful treat in days gone by. Ice cream making was a ritual starting with the ice harvest as Dad has described. I have no recollections of the ice harvest itself, but I do remember as a child going down into the ice house with Dad on a warm summer day, digging through the straw for a piece of ice, and putting it in a gunny sack. Then out on the lawn we would smash up the ice in the gunnysack by hitting it with a hammer. Mother would have the canister of ingredients ready, which was put into the crank ice cream freezer, the crushed ice and rock salt was packed around the canister. During the freezing process the excess salty brine, which accumulated during the cranking, drained out of a little hole near the top of the wooden freezer. I think the freezer was set into a big granite dishpan to catch the brine, as we were always out on the lawn. We kids got to take turns cranking, until we could crank no more as the ice cream became frozen. At that point mother took over again, pulled out the canister and opened it up, everyone watching in anticipation. Our first taste of this special treat was when she pulled the dasher from the canister, placed it on a big platter, and we all had our spoons waiting for that first bite! The canister was then replaced in the ice/brine and this would keep the ice cream frozen until time for dessert. As I remember, ice cream making was always done outside and often when we had cousins visiting. A great summertime activity and lots of kids to do the cranking of the freezer!"

WHITLASH HITCHIN' RACK

Whitlash used to have a hitching rack to tie teams or saddle horses to. It was a heavy steel oil-well cable from post to post, stretched pretty tight and high. This bronc bucked and run his head under the cable, and lucky me, the cable caught the saddlehorn and stopped him. A wee bit higher, "Oh, Oh! My poor belly button!" Just this year I looked at that saddle and the marks from the cable are still real plain on the nickel saddlehorn. Jon Woods is now the owner of that saddle.

"Old" Whitlash on Breed Creek, Sunny Side School on hill—1905

Whitlash—1905 1) Creamery, 2) Saloon, 3) Hotel, 4) Store, 5) Blacksmith Shop

Burnham Murray Store—circa 1950, Whitlash present location

Whitlash rodeo, 1920s

WHITLASH MAIL SERVICE

Many years ago our mailman delivered the mail with a team of horses and sled in the winter. He left Chester and went to the Hill, Montana Post Office. Another team and sled with mail from Gold Butte first went to the Whitlash Post Office and then on to Hill. Both drivers changed to fresh teams there. One headed back to Chester, and the other driver went back to Whitlash, and then the nine miles to Gold Butte. Later, in the late 1930's, we had a mailman who came to Whitlash in a Model A Ford equipped with caterpillar tracks and skis, which I described in a previous story.

This was before four-wheel drive pickups and snowmobiles were invented or thought of. We didn't have graded or any graveled roads then. During all those years our service was six days a week. Now its three days a week on graveled all weather roads with a four-wheel drive!

Well anyway, that's often enough to get bills!

WOLVES

Wolves, killing our cattle, were a big problem when I was a kid. At that time there were no antelope, deer, or elk in the Sweet Grass Hills, so beef was their diet. They could outrun a cow critter and hamstring them easy. Now we do have the wild meat, but they are a lot harder for the wolves to catch than a cow. I was with my Dad one day and we rode over a hill and here was a big gray wolf with one hindquarter already eaten, and the cow lay there still alive.

At the time Lester Stott on Breed Creek was milking cows for a living, and the wolves were eating up their cows. Soon there was no butter to sell or cream to be taken to the Whitlash Creamery.

John Oswood, a Government trapper and Joe Gardipee finally shot and trapped what wolves were left in the Sweet Grass Hills. I believe that was in 1927. The last summer the wolves were here, Mr. Strode had a cowboy go out at dark and ride the hills and shoot off a rifle real often to scare them over the next hill.

Steve Boyce who was in the same age bracket as my Dad was a good friend of ours. He was a Government hunter from 1924 to 1937 in

Valley and Phillips counties, (and maybe more, too) and trapped and shot a lot of wolves. They said he was tough on their trail with his saddle horse and packhorse, and he spent a lot of time in severe weather just living on these trails. I asked Steve years later how come he started trapping and shooting wolves. His answer was this. "When I was a little kid a wolf came on our ranch and killed a little colt whose mother was my pet saddle mare. From that day on I hated them and promised myself when I got grown up, I was going to kill all the wolves I could find in the States and Canada."

And now these educated men say, "On, no, these beautiful wolves won't kill cattle". Maybe these guys figure wolves eat grass.

John Oswood and wolves

RATTLESNAKES

The Old Timers were always proud to tell a Newcomer that The Sweet Grass Hills had no rattlesnakes. But in 1940 they started migrating over the Bingham Ridge. Beyond the Bingham Ridge on up into Canada there had always been rattlers.

They were plenty thick that first year around Iversons and Schafers and where Jimmie Morrison lived. The next year they were around the Whitlash area. Within a few years they were into Middle Butte and West Butte. However, as the years went by there were less of them around, just enough of them to keep you real snake cautious. We still see a few every summer. They seem to be migrating and all going in a southwest direction.

Elsie and I were following a loaded truck on the road over by Dores Creek and it ran over a big rattler killing it and the snake was full of Sarvis Berries. Byron, my nephew was telling a snake expect about this, and the expert says, "You tell your uncle that rattlesnakes just don't eat berries." Many of these berries grow on bushes not over a foot off the ground. Anyhow, seeing is believing and I hadn't just walked out of "The Long Branch" having been there too long and "seeing snakes"! From that day on, Elsie was none too keen on picking Sarvis Berries, either!

WINTER ENTERTAINMENT

In the days before TV and all weather roads, the ranchers played a lot of cards during the wintertime. Every Saturday night, weather and roads permitting, there would be a card party. The games played were either pinochle or 500. Often there would be seven or eight tables set up wherever there was room, from the living room, to kitchen and or dining room and in the bedrooms. It was also a potluck, which was served at the end of the card playing. People would take turns from one ranch to the next, hosting the party. Prizes were given to the man and to the woman with the highest scores, and also for the lowest scores, the "booby prizes". Before we had antifreeze for the car radiators, and if it was real cold, everyone drained out the water, and filled it up again when ready to go home. The other option was to keep the motor running with a horse blanket covering the hood of the car during the whole evening. The temperatures in the winter were often 30 degrees below. They were great get-togethers!

The community also put on many Home Talent Plays at Liberty Hall. Playbooks were ordered by mail and given out to the cast so the actors and actresses could learn their lines. My Mother was most always their director. There was a lot of great talent out here in the "boondocks" in the Sweet Grass Hills and everyone had a lot of fun. When the play was perfected, it would be performed at Liberty Hall, and the Hall received the ticket sale proceeds which were used for maintaining the building. Many people from neighboring communities came to see these plays at Whitlash. Some winters the play went "on the road" and was shown in some of the High Line towns.

As a teen-ager, I'd go up to the Hall and start a big fire with wood and coal in the big pot-bellied heater, so as to get the stage and building warmed up before the practice sessions. Our family often took this responsibility, as we lived less than a mile away. When I got a little older, I also took a part in some of the productions. These plays were great entertainment for both those performing and the audiences who came to see them.

Cast of Home Talent Play—Whitlash
Standing: *Marion Parsell, Harry Demarest, Harold Brown*
Sitting: *Vera Brown, Myrtle Johnson, Babe Furnell, Louisa Stott Allen, Murray Johnson, Myrtle Leach, Ted Allen*

SCHOOLS OF THE COMMUNITY

Just a little on school houses in this area. The first school, called Pleasant View, District #27 of Choteau County, was held in a log cabin west of old Whitlash in 1899. I don't know for sure the exact location, but I believe it was by the Fitzpatrick Lake where at one time or other was the location of the Whitlash post office. The teacher was Fanny S. DeLacy. C. J. Keller, W. A. Carrell, and George Murray were the members of the school board. The pupils were Agnes Carrell, Mamie Carrell, Loie Murray, Lena Murray, Lottie Amez-Droz, Elmer Brown, Esther Brown, Harold Brown, Amelia Brown, Edna Ellis, Myrtle Ellis, Mary Fitzpatrick, Lynda Rock, Charlie Farrell, Jessie Rowe, Ada N. Rowe, and Edwin Sommers.

John Brown and Lester Stott built a school building located in the southeast corner of Section 36 next to the Frank Harrington Homestead, also not too far north of the Whitlash Cemetery (which at that time wasn't there) and it was named Pleasant View School. It was built around 1904 and the pupils were Alice Stratton, Lois and Lena Murray, Bessie Brown, Amelia Brown, John Stratton, Archie Stratton, Howard Stratton, Johnny Morgan, Letha Morgan, Mary Morgan and Golden Fitzpatrick. I think the school was only at this location for a couple of years and then was on the hill west of the first Whitlash town site. At this location it was called The Sunny Side School. This is where my three sisters and I went to school in later years.

The Sunny Side School was moved to the Bob Thompson Ranch and called The Thompson School in 1933 with Elsie Wehr being the teacher. Later on it was moved into Whitlash where the church manse is now located, and then it was called The Whitlash School. Auverne and Doug attended it at that location. The next move was south of Lybeck's and Woods and was used for their families. Later on after bus service started out of Chester it was then moved to the Woods residence and used for a garage and workshop. Three generations of Demarests went to school in this building at four different locations. After the building was moved to the Lybeck area, Whitlash District #27 got a new building. At present we still have a country school at Whitlash. A bus comes up from Chester to take the high school scholars to the Chester High School.

The North school was moved to five different locations to accommodate different families. This I'm not sure, but the first location of the North School was between the Johnson Ranch and Fromm Ranch next to the division fence. That would be the location where the Murray Johnson kids went. The next location of the North School was south of Gardeners, next to the fence between them and Parsells. The third location was on the Hill where David Parsell's road goes to his ranch, and then it was moved east of Gardener's down towards the Sutton Ranch. The last location of the school was at the crossroads west of the Brown Ranch. The last move of that building was to Whitlash, but school wasn't held there. It was taken down piece by piece by Buck Steinback and used to build a garage and a couple of calf shelters, and a kid's playhouse. Elsie Wehr taught one term in this schoolhouse when it was over by the Harry Gardener home in 1931. She boarded at the Elmer Brown home. Two of her pupils were Irvin Brown and June Brown, and they drove a one-horse buggy to school.

In the spring of 1912 a school was moved from Strawberry Creek area to a location north of where Bobby Thompson's corrals now are in his field on the west side of the lane pretty close in the corner of his field to the south. The markings in the soil still show up real plain. The Stott, Hicks, Emelines, Robinsons, Larsons were some of the children who went there. Some time later it was moved west on the Strawberry Road just a ways south of Ron and Toni Wehr's road leading to their ranch. Matt and Frances Furnell were among the pupils there. Then in 1930 it was moved to a location a few miles south of where the present Richard Wickum & Sons Ranch is today. Later it was moved to the Lloyd Oswood Ranch and made into a shop. To my knowledge it was called The Strawberry School while in our area.

The Bear Gulch School, District #8 is where the Bingham, Iverson and Townsend, and maybe the Price youngsters attended school. These "youngsters" I refer to were of my generation. The first location was on a hill on the west side of Dores Creek just a little northwest of where the present county road crosses Dores Creek. Later the building was moved to the Whitlash area by a neighbor who bought it for a granary.

A new school was built and put on a good cement foundation a little further south of the Iverson grain bins. Elsie Demarest taught several terms there; 1936,1943 and 1944. Pupils in 1943 were our daughter, Auverne who was in the first grade, Larry Roberts and Lyle Roberts, and Barbara King. In 1944 the pupils were Auverne, Carol Bingham, Punch Rambo and his younger brother for part of the year. The Rambos were living on the Strode Ranch, but moved away after school started, so most of the year Elsie just had Auverne in the second grade, and Carol Ann Bingham who was in the first grade. Auverne was the only one in her class through all eight grades of grade school. Douglas was not yet school age, but he stayed there part of the time with Elsie and Auverne who stayed all week. They usually only came back to the ranch on weekends. At that time eight miles between our ranch and the school was a long way to drive over the bad roads and deep snow. Another fact being that there was only one car on the ranch then, and it was before we had any pickups. Later that school building burned to the ground and a new school was built at the crossroad north of the Schafer Ranch on the north side of the road. Elsie also taught here a short time until they could get a new teacher. Then it was moved up the road south from where the other building had burned close to Bear Gulch. Again, the building was moved to Whitlash and used to take care of the overflow of pupils at the Whitlash School and still remains there.

As you can see, the above schools were moved multiple times to different locations. They were pulled on big log skids or timbers with horses or old time tractors. Only the school that burned down had been put on a cement foundation.

In the 1920's, 30's and 40's very few school terms were for nine months. School was held until the money set aside for education was gone. A lot of six, seven and eight month terms were all the pupils got. Nowadays if money runs short the county votes a levy for more funds. But it was different back then. When the school money was gone, there was no more money for a levy either, and that was the end of that school term.

School Picnic 1924—Anna Iverson, Oscar Iverson, Ellen Stott, Louisa Stott, Bill Townsend, Ray Stott, Myron Stratton, Mrs. Stratton, Harry Demarest, John Oswood; Carl Iverson, Ruth Stott, Harold Brown, Unknown lady, Unknown child, Ruth Stratton

School Picnic 1924—Ellen Stott, Ruth Stratton, Ruth Stott, Mabel Demarest, Carl Iverson, Charlie Howery, Jimmy Townsend, Nora Townsend, Annie Iverson, Unknown, Unknown, Claudie Demarest, Dot Thompson, Helen Demarest

COUNTRY SCHOOL STATE EXAMS

When I was in grade school, the country school pupils had to take a state examination. We took tests on all the subjects, and if you didn't get a passing grade in a certain subject, you had to attend school the next year or said subject until you passed it. The state tests for agriculture and physiology were given in the seventh grade. All the other subjects were given in the eighth grade year and no going into high school until all subjects were passed. Why the country school pupils had to take these state examinations and the pupils who lived in town weren't required to take them, I don't know. However, it was said that the pupils who had to take the State Exams were by far more prepared for high school than the town kids. Many of our country school students were at the top of their high school classes. That is except Claudie. He passed those State Exams, but then when he went in the front door of the Chester High School, he ran right down the back fire escape, and headed home for the Hills and his horses.

A certain day was chosen for State Exams and the exam was held in your own schoolhouse. A sealed envelope was mailed out to the person who was chosen to be with the pupil while writing their test. My mother, Amelia Demarest, was in charge of the pupils and their written tests for many of the kids in our area. She volunteered to do this and I guess they trusted her honesty and ability.

THE WHITLASH DANCES

The community built Liberty Hall in 1917. For about thirty years Elsie and I were the managers. Up until the time Whitlash built a church in the 1950's, all of the church services, funerals, weddings, and any get-togethers were held at Liberty Hall. During the summers and falls we held dances every two weeks. We sold dance tickets at the door for one dollar a couple.

Each member of the Ladies Club made a salad and baked a cake, which they donated for the supper that was served for fifty cents a plate. The Hall provided the bread, meat and coffee. The Ladies made the sandwiches and coffee, served the food buffet style in the Hall basement about midnight and cleaned up afterwards. You were always glad when your partner was not one of the ladies on the supper committee for the evening, 'cause she didn't have much time for dancing! When the supper waltz was played, that was your clue to find your wife or sweetheart, dance a round or two and then head for the basement for chow. We always had more dancers than the floor would hold. It's a good thing there was lots of country outside.

Many Canadian couples and singles were there, and after "Home Sweet Home" was played, they all left at the same time to make it easier for the Custom Man to get them herded back into King George Country. These Canadian boys were quite popular at our dances when the U. S. was "dry" except for moonshine whiskey or good whiskey bootlegged from Canada. The Canadian neighbors always had some Corbey's Special, which was "as smooth as lemonade" to treat some of us Yankees, and the Corbey's Special never lasted till morning!

For several years we had an orchestra from Canada, The Mueller Brothers. Other musicians were Renie Turner and his wife Marian from Ferdig, Montana; and also Stanley Bunyak, Johnny Dolezal, Cliff Hanson, Eddie Ross, B. J. Oswood, and B.J's father-in-law, Don Gummer. The orchestra usually was composed of a piano player, drums and guitar and saxophone.

One of my responsibilities was to keep the drinking and rough goings on out of the dancehall, and now and then things did get a little "Western"! Actually I had been deputized by the Liberty County Sheriff and carried a silver star Deputy Badge in my pocket, but I never wanted to show it and I never did. For years a Deputy or Sheriff came to our dances, but when Tiber Dam was being built, they had to stick around Chester and keep order in their own town on Saturday nights.

These dances usually started about nine in the evening, and ended at two a.m. Kids were there too, and learned to dance at a young age. Later in the evening the little ones would be tucked in to sleep on a bench somewhere. In addition to the regular dancing, every so often a square dance would be called. Sometimes when the crowd was having a great time, they would pass the hat for more money to pay the orchestra, and then there would be an hour's more dancing. It was not unusual in summer for daylight to be breaking by the time the last dancers left and the orchestra got packed up. Then we stayed and swept out the Hall, rearranged the chairs that lined the walls of the room into two rows of chairs with a center aisle, ready for Sunday morning church. The pulpit and altar would be brought out from the little storage room where all the "church things" were kept, and Liberty Hall was transformed from a dancehall into a church sanctuary.

In later years my daughter remembers going up there in the morning with my mother, Amelia, before church to get the grounds cleaned up. My mother had Auverne and Doug helping her pick up all the whiskey and beer bottles littering the grounds around the Hall from the night before. Then she would place two big bouquets of fresh cut flowers from her yard on the pedestals on each side of the altar, and the place was truly ready for the eleven o'clock worship services.

When Elsie was once asked about some of her favorite memories, she replied, "The Whitlash Dances. Every third dance was Claude's and mine!"

IX / Crossed Paths Along the Way

THE DRIFTER FROM WYOMING

We had a cowboy working for us by the name of Bill. He drifted in from Wyoming with a bedroll and saddle. For two years he worked between our ranch and the Brown ranch and for whoever needed help.

In 1921 Dad and some of his neighbors bought some yearling heifers from someone in Utah. They were shipped by rail to Galata, Montana. After they unloaded these heifers they were heeled and branded. Each rancher branded the appropriate number with his brand, according to how many of the bunch he had purchased. I remember some of these heifers had a , a Utah brand, and some had a 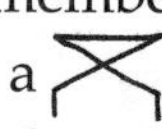. I was always asking Dad, "What do you call that brand?" In order to satisfy his "brand crazy" kid, Dad answered, "We will call it a "hootenanny". Of course I then wanted to know what that was. He gave some answer, but I don't remember now what it was.

Dad and the other ranchers then trailed these heifers from Galata to the Sweet Grass Hills. They were only a couple of miles from our ranch when a car pulled up and two strangers asked Dad if a man by the name of Bill _______ was working for us. Dad pointed out which cowboy he was. They, "The Law", stepped out of their car and pulled a gun on Bill and handcuffed him. He was suspected of murder in Wyoming. When they got to our ranch, Dad told them they had to take the handcuffs off him as we kids thought the world of Bill and just wouldn't quite understand. Bill was uncuffed, went in had dinner and Dad settled up his wages with him. On leaving, Bill gave Dad his 38 Smith

& Wesson six shooter, saying "Looks like I might not need this for awhile." He also gave Dad a pair of angora saddle pockets. I can just vaguely remember when he left. He always made a fuss over us kids.

THE FUNERAL OF ISABELLE SUTTON

In December of 1929, word had come from Kalispell that Isabelle Sutton was very sick. She had been going to school there. Her brother, Robert Sutton and Jack Taylor, who was her beau at the time, left in the late afternoon on horseback for Chester. I remember they stopped here and Dad gave Robert one of our horses to ride. They rode all night getting there. Their plan was to catch the train for Kalispell, but a telegram was waiting for them at the depot, saying she had passed away. Her body was brought to Chester for burial, as her father, Frank Sutton was laid to rest in Chester. Three of us local boys, George Strode, Verland Taylor and I were asked to be pallbearers for her funeral in Chester. It was way below zero and the snow was deep and crusted. It was a tough ride on saddle horses and about forty miles. It took all day to get there. Her mother and Bob Robinson and family went with a team and sled to Chester. At the cemetery the next day it was teams and sleds (no vehicles). We came home the third day. It was a real sad time for our community. I often remember this experience on occasions when I think to myself, "Am I going to have time to go to so-and so's funeral today?" I don't have to think too long about it, and I go.

THE T. P. STRODE RANCH

When I was a teenager, my Dad always helped Mr. Strode whenever he gathered and worked cattle. He helped with gathering for branding, then getting out the yearling heifers to run separate from the bull. At that time everyone bred twos to calve at three instead of two year olds. Mr. Strode made two shipments every fall to the Corn Patch, then there was more cow punching to be done at calf weaning time.

After I got a little older Dad didn't go and help, but he sent me. "Happy Me", I helped every season till there was no more T. P. Strode Ranch. At that time it was one big field from Breed Creek taking in Sage Creek. One day we would gather Breed Creek area and the next day gather the Sage Creek area. This of course was the high country, not the winter range. So I know that ranch like the back of my hand. Mrs. Strode was such a gracious lady and was really good to me. I always wore bright colored shirts if I could, and she was great for color, and always bragged up my shirts. In later years I helped trail the last beef shipment from the Strode Ranch to Chester. They were put on cattle cars to go to South Saint Paul. When pushing the steers up the chute Mr. Strode remarked, "When these are sold, I'll be out of debt again". The next day he went to Havre and passed away on the street. Edgar Woods went with the shipment to the Corn Patch, as at that time he was working on the Stirrup Ranch. Mr. Strode always rode when we gathered cattle and when shipping, he always went as far as Deer Creek at the east end of his ranch.

Mr. Strode took a livestock magazine named "Drovers Journal". I do remember looking at it and it was quite a thrill to me because I had never seen a magazine about cattle and supplies before. There may have been more livestock publications, but as a teenager I looked forward to thumbing through the pages of livestock pictures and stories. At this date our ranch takes the magazine. "Drovers" was founded January 11, 1873, so in this year of 2001 that makes the "Drovers" 128 years in publication.

EDGAR WOODS GETS MARRIED

When Edgar Woods was on the Strode ranch in the late 1930's, I asked him when he and Dolly were going to get married. "Tomorrow", he says, "if I had somebody to feed the cows." "No problem," I say, "get going and I'll come over and stay and feed till you get back. Just don't make the honeymoon over a couple of days". At that time loose hay was fed. The Strode Ranch used a high-wheeled wagon equipped with a hayrack. I went over to Dores Creek where the hay was stacked to load the wagon. It took most of the day to feed those cattle.

Anyhow, the evening before I looked for them back, I cut a lot of long white tail hair from a white horse down at the barn. I cut this hair up fine, about a pint jar full, and sprinkled it real thin all over the white sheet in their bed. Then I collected all the alarm clocks on the ranch and hid them all through their bedroom, set for different times to ring in the wee hours of the morning. Then I gets a box of Cornflakes and scatters them from the door of the bedroom clear across the room to the night stand where they had to light the coal oil lamp. I heard them come up the stairs (my room was just across from theirs). You can't imagine all the noise walking through those Cornflakes! All the way across the room, Edgar was hollering, "God Damn that Demarest! God Damn that Demarest!" I quickly made myself scarce and went down to the barn, saddled up and lined out for home!

When Elsie was teaching down at the Hawk's School, I made the remark one weekend, Why don't you make me five different kinds of pies for next week while you are gone? She did just that. Edgar and Dollie Woods stopped in one day and said, "We dropped in for a piece of pie", knowing that I was batching and was not a pie baker. Says I, "What kind do you want?" You ought to have seen the surprised looks on their faces when this cowboy served them each just the kind of pie they ordered!

JOHN, SAM AND HARRY MAKE GASOLINE

During the late 1930's and into the '40s, a bachelor neighbor John Stratton and his friend Sam Watson refined crude oil into gasoline. Sam was staying with John at the time. They sold it to a lot of neighbors to burn in their cars. They got crude oil in barrels from the first oil wells discovered in the Sweet Grass Hills on the Hick's lease. Their refining process sounds similar to the way people used to make bootleg whiskey out of grain, berries, dead dogs and cats in prohibition days. Anyhow, when they were brewin' oil to make gasoline there was a big cloud of smoke in the air coming from their direction. I remember there was also quite a lot of smoke coming out of the exhaust pipes in those cars using this gasoline. But it got them down the road!

A few years later our area north of us had a lot of natural gas wells. When they blew these wells to get the moisture out of the gas, in some way they were able to catch this moisture, which was called "drip gas". At that time Harry Gardener was working for Montana Dakota Utilities, and he used this drip gas in his car and said it was exceptionally good. It got them from here to there and was free, too.

This same crude oil was burned straight in our Canadian neighbor's tractors. Many barrels of oil in old time trucks went by our ranch to Canada. I don't know what it cost a barrel, but it must have been pretty reasonable. I'm sure the exhaust smoke drifted clear to the Cypress Hills. At this time our ranch was still using hay and oat burners and "cracking the whip" for more power.

Remembering Sam Watson reminds me of something he told me one day. I had a bronc and a gentle horse hooked to a two-wheel "dead axle" breaking cart that had a place on the back to stand. Elsie was with me and we headed for Whitlash with this bronc half buckin' and running away. When we finally got stopped at the city, Elsie jumped off and went into the post office to get the mail. About that time Sam came up and says, "No bloody well wonder you folks don't have any kids yet, riding your wife around in that bloody outfit." Sam was an Englishman and really had an English brogue. They "lost the mold" when they made him; he was quite a character. One day I was riding down the road horseback and was "buckin" a strong wind. I saw Sam walking down the road too, in front of me, his 110-pound frame all humped up. On catchin' up with him I slapped him on the shoulders and hollered and really startled him! He really gave me a good cussin' out in that English tongue, and called me lots of words I didn't know were around.

The breaking cart—Steve Babokus, Claudie, 1935

DENNIS RAPP

I became acquainted with George Rapp when I was with cattle in South Saint Paul. He weighed the cattle at one of the scales. He had a son, Dennis Rapp who spent three summers on the Demarest Ranch. The first time Dennis came out he was about 16 years old. Auverne was about eight years old, and knew that he was coming on the train one evening. My dad went to Chester to pick him up, and he was put to bed in an upstairs bedroom at Dad's house. The next morning Auverne was curious about this kid from St. Paul and went over to see him. Well, he was still sleeping, and her Grandmother Amelia said she could go quietly upstairs and peek at him, but don't wake him. Well, what she remembers is that she really couldn't get a good look at him all wrapped up in the covers, but she vividly remembers the *BRIGHT RED* cowboy boots by the side of the bed. She was used to seeing lots of boots, but never ever a red pair!

He turned out to be a good hand with a horse and also a good cowboy. In the fall he went back of the Hick's place and came out with a big mule buck deer. Dennis fit right in the community at get-togethers and with the 4-H club. He played an accordion and was frequently requested to entertain at home and community events. Everyone loved his accordion playing. He was very interested in forestry, plants, trees and grasses and went on lots of trips into the Hills with the 4-H groups helping them identify the trees by their leaves or needles, the bark and the seed pods.

In Dennis's mind he figured some day to be a rancher so he sent to the Montana brand office and got a livestock brand 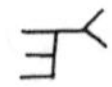 recorded in his name. Auverne and Doug had a few calves and he bought a heifer from them. We have a picture of him putting his new brand on his start in the cattle business. He got to help us with a beef cattle drive to Chester one fall. Elsie followed with the cookie wagon and Auverne and Doug, Dennis and myself were the cowpunchers. We had two nights of night herding on that trip. It was a great trip and a little frosty the last night out. Elsie sat up all night keeping a campfire going for the nightriders.

On another occasion I remember we were gathering cattle next to the road at the Murray Place and Dennis's horse fell with him. Finally

I noticed him sitting on a big rock with his sock and boot off trying to wiggle his big toe, and a "little green around the gills" and cussin' quite a little bit. He finished out the day with a naked foot dangling next to his stirrup and his boot and dirty ole sock tied on the saddle. On getting home I wanted to put his toe in a little cast, but he wouldn't go for it, especially because of me being just a cowboy vet. Every time I go down the Strawberry road yet today, I can see Dennis sitting on that big rock in the field next to the road, holding his big toe and I can't help but grin a little.

One fall Dennis after he had graduated from high school, he was going to winter with us and we found out he had won a scholarship to help him out in college. He got the scholarship for being the champion broad-jumper of Minnesota. So we just loaded him in the car and put him on the train for home. And I told him, "You just get all the education you can, then think about your future."

I just remembered I was breaking a sorrel mare by the name of Lassie. She was a beautiful mare with a bald face and had four socks. I told Dennis one evening that he better start riding her so she would get used to a different cowboy mounted. So he stepped on her to go get the milk cows. On coming back he told me that when he got out in the field and next to a fence, "she just grabbed her butt, and bucks me off across the fence, and me in one field and she in the other". I kind of grinned at the time, but do you know I rode her for several years and she shed me several times. She was a good cow horse too, but never did forget the habit. In fact I finally rode her full time with a "cheater" which is a piece of clothesline rope placed firmly in her upper lip and attached to my bridle bit. I guess it put pressure on a nerve and a horse wouldn't take their head to buck. She was a good cow horse, and I also heeled calves on her.

Dennis was asked to be in the "Crows Nest" with the rodeo announcer and timers, in Sweet Grass, Montana and play the national anthem on his accordion at the start of the performances. While he played the "Star Spangled Banner" everyone had their hats off, but Dennis! He also played some good "toe tappin' music" between events that afternoon. He did a beautiful job. The Beer Barrel Polka was a big hit on his accordion; he could really get you all fired up with that tune, and Doug was after him to play it all the time. Part of one summer Dennis's sister, Arlene, was on the ranch with us.

Dennis completed his college education and later worked for the Forestry Department. He and some others were sent one summer into far northern Canada by plane to do some research. It was an area where men had never been. Later he worked full time with the United States Department of Commerce dealing with other countries and he was in about every country in the world. We are proud of you Dennis in your career; but damn happy you were first a cowboy who owned a recorded Montana brand and a few cattle in the Sweet Grass Hills!

Dennis branding his calf, Auverne and Claudie—1949

Dennis with mule deer

SID HOUSE

Sid House worked on our ranch for a few years in the 1950's, and I have a few stories to share of some of his and my experiences. On this occasion Sid and I went over to help Dolezals gather cattle. We were riding next to Deer Creek on the west side and I didn't see the "wreck", but I guess Gidget fell down the side of the mountain with Sid. You know, I still think she bucked him off! Anyhow, the horn of his saddle caught in the "fly" of his pants and ripped them off him right down to the top of his Tony Lamas. When I came in sight of Sid he was tying what was left of his Levi's on the back of his saddle. All he had left on were the belt loops and his belt. But lucky Sid! That morning he had put on an extra pair of pants. It was a good thing, 'cause it was pretty nippy out that day, and punchin' cows the rest of the day in his shorts, — well he may never have had any grandkids.

One evening Sid and I were riding not far from home, and Alberta took a bite at Sid's horse and I jabbed her with a spur for being naughty. She really blew the cork! I lost one stirrup and then the other (of course my excuse was that they were wide tread roping stirrups). Anyhow, I took up another homestead. What bothered me most was that Doug had been riding her and getting along real good. And I had to go and mess her up. I had Sid go back to the barn and get my Oxbow Stirrups and hoped I could get the job done with them. So I changed stirrups and I was in luck. She never offered to drop her head again.

Sid and his new Hamley saddle in our living room. It was shipped in a burlap sack from Pendleton, Oregon

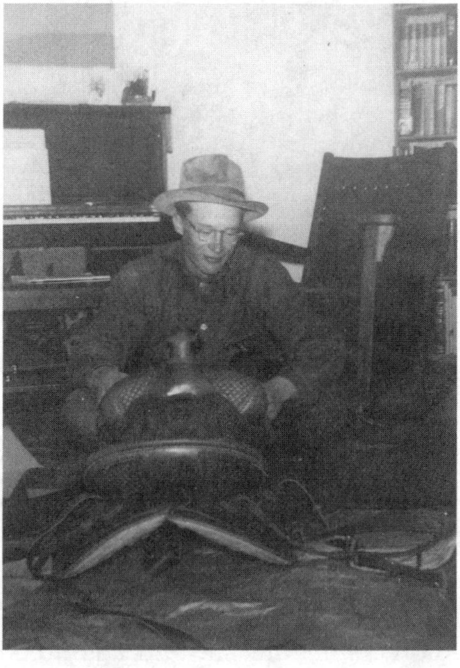

Another time Sid and I were riding along and he was riding Compass. He was breaking her to ride, but we were going along side by side and talking. I looked off in the distance for a moment and then on looking back to visit some more, Compass was along side of me, but the Hamley was empty. Sid said he "got off to pick a pretty flower, and she left me". The only thing is that I noticed he was a little "gimpy" on his left hind leg!

Our creek over at the calving shed was in flood stage, about twenty feet across and about belly deep to a horse. Nickie was sharp shod in front, but not behind. There was ice going into the water, but the water current had melted the ice where it was swift. When Nickie's hind feet hit the ice he tipped over backwards on me. I was lucky my head was still out of the water and the swift current of the water took me to shore. Sid House was with me. I had just had a cold, cold Saturday night bath and no towel and a wet saddle.

HIGH MOUNTAIN TRAILS

A good friend of ours, Delbert Badberg and his daughter, Barb, came one summer from Nebraska to visit. They rode a lot of horseback miles in the Hills and had a good time. Another time Delbert and Doug were staying at the Hicks cabin with the Nebraska cowboy cookin' the flapjacks one morning. He gave Doug one and then gave one to Cindy, our big black Labrador dog. She grabbed it and then kind of hesitated, looked up at Del, and dropped it on the cabin floor and went over in the corner and lay down. Del looked at Doug kind of bewildered and beat, and said, "You ate yours, didn't you?"

Another year Delbert's dad, Harm came out to see us. One day we put some block cattle salt in this 1948 International pickup and drove away up high in East Butte. After getting up there we were on a narrow trail and below us was our salt ground down a hill, steep as a cow's face and about six town blocks long. I remarked we were going to put out some salt down there. Harm said, "Oh, no, I can see from here that you got plenty left!" The sound of his voice caused us to take a look at him, and he was white as a white Leg Horn hen, and he had his fingers hooked through a little crank on the windshield with a death grip and staring straight ahead. His poor fingers looked like they hadn't seen any blood flow for some time. We three hadn't realized what he had been going through. So we kind of quit our carefree BS and got Harm kind of back to normal and finally came to the salt lick which the mountain trail went right by. But I don't think Harm was fully recovered until we hit the low country. We had a great time while they were on the Ranch and their visit was a way too short, like a buffalo's tail.

Mother, Dad and I were traveling back from a trip to Nebraska. We went into the big Horn Mountains from the east side and it was just switchbacks and very narrow, really a tough road all the way to the top. On top was a big plateau and it was pretty level. Dad says, "I'm sure glad we are off this mountain." This was in July and I say to dad, "Look we are just on top. See all this snow in the ditches". Anyhow, going off the west side was really bad. Dad says, "When we get to the bottom of this darn mountain, I want you to stop the car so I can get out and kiss the ground." And that's just exactly what he did! We visited and stayed with old friends, Jim Leach and his wife Myrtle in Lovall, Wyoming that night.

DAVE AND DONNA BRYAN

Stu Smith was on a Stockman Tour in the Corn Patch and feedlots around Sioux City and got acquainted with Dave Bryan who worked as a cattle buyer for Producers Commission Firm. The next year Dave and his wife Donna Bryan came out in our area and got acquainted with all the ranchers and order bought a lot of calves for feeders in his area. They started coming out our way every year and helped many of the ranches at branding time in May. Dave likes to rope and heel calves, and believe it or not, once in a great while he does double hock an odd calf! (Ha, ha!) just foolin' Dave!

Dave tells about one time he was at the Les Eide Ranch and getting ready to go look at their cattle that were a couple of miles away. It was about 100 degrees in the shade and just as they were starting out, Lester Eide (Les's Dad) said, "Wait! I got to get my sheepskin coat!" You can never trust the weather in Big Sky Montana. The saying goes, "If you don't like the weather, just wait a bit!"

You bet, they are a great couple, that is if you aren't too particular whom you associate with! They had a nice little ranch in Iowa. I visited there once and Dave drove me around and I met a lot of people and feed lot cattle. But as for Donna, as soon as I showed up, I guess she didn't want to put up with me and she got sick and went to the hospital.

Dave and Donna did our ranchers in our area lots of good, buying calves for the feeders in the Corn Patch, and we had some fun times. Their brand is ◇X. Dave is also an auctioneer, and comes out sometimes in the fall to auction at Buster and Helen Brown's horse sale. They

Dave Bryan, Les Eide, Claudie, Lester Eide, Doug—1994

sell weaner colts and some broke saddle horses at the Fair Grounds in Shelby, Montana.

Though Dave and Donna have moved to Missouri, they still make it up to the Sweet Grass Hills now and then at branding time in May. They were here for the year 2000 brandings and everyone in the community is always happy to see them again.

BYRON, BUTTERFLIES AND RHUBARB PIE

Byron Weber, my nephew, has been coming up to the ranch since he was a very little boy. He used to help my mother (his grandmother) with chores around the yard, and picking chickens. She always said that he was her best worker. When he got older he worked summers helping us fence, hay and stacking bales. Byron is also a naturalist and an expert in the identification of plants, animals, birds, and other creatures. Now he brings his own little boy, Nicholas, and comes to visit and catch butterflies. He spends quite a few days each year in the Sweet Grass Hills hiking and catching butterflies in different seasons. No one else has ever before studied and documented the different species of butterflies in the Sweet Grass Hills. Of an evening he has several hours of bookwork to categorize the kinds and exact location where each one was captured. He also deals and trades with other butterfly people. He teaches school in a little town south of Missoula, Montana.

A couple of years ago for branding Byron even brought two homemade rhubarb pies, which he had "made from scratch" himself! We're still wondering how he knew that our branding cook was disappointed not to find any rhubarb at the ranch that was big enough for pie baking. It is traditional that a variety of pies be served to the branding crew, and a rhubarb one is always included. So she was grinnin' when she saw those pies were *rhubarb* pies, and boy, were they good!

Byron left a butterfly catcher with Doug and me. It has a long handle to hold on to and a little open sack on the end to use in case we see any butterflies. Anyhow, when Doug and I are on horseback, both with a coil of grass, it just seems like its hard to dab a small enough loop on the little flying rascals and take up the slack to catch them, and soon a thirty foot of hemp is way too short. Sorry Byron, but we cowboys just can't seem to get the hang of catchin'butterflies!

MA'S MILK COW

The day before Christmas, half a foot of new snow,
Winter in Montana, got four months to go.
Half asleep in my chair, watching the games,
Edmonton Oilers play the Calgary Flames.
Ma comes in from the kitchen, bucket in hand,
"Need milk for the fudge, Love, Darigold brand."
Well, no dairy would have her, she's no blue ribbon winner,
Got a bull calf beside her sucking down dinner.
She lets out a beller, a little nervous right now,
Darigold, they call her, Ma's milk cow.
One broken horn, double hump in her back,
"Stand still you old heifer, I'll give you a smack."
Not a tooth in her head, a worthless old rip,

Walks pretty funny, down in one hip.
There's one faucet working, doesn't feel like a lot,
"This is for Ma, now, give what you got."
No cream on the top, its all half and half,
I figured these squeez'ins belonged in that calf.
Not much in the bucket, a tough one to crank,
Tease her a little, a couple more yanks.
I finish her up, her tail gives a flick,
She steps in the bucket, and gives it a kick.
Flat on my back, I land with a thud,
Covered all over, manure, milk and mud.
"Ma can go without," I cuss this old sister,
No Christmas fudge milk from this swing-bag blister.
Throw open the gate, slam shut the barn door,
Jump in the truck, and head for the store.
I march in the house, and throw on the table,
Two pints of canned cow, just look at the label.
CARNATIONS'S the name, it's the best in the land,
"Here's your damn milk, Ma, right in the can!"
"There's no teats to pull and no hay to pitch,
Just punch a hole in the son-of-a-______...!"

by Lon Hoppe

LON HOPPE

Lon Hoppe is a friend of Doug's and mine. He has often been part of our branding crew. Next time you're heading north from Shelby, a few miles south of Four Corners on the east side of Interstate # 15 you can see his elk. At least that was where he was the last time I saw him. Sometimes that elk gets tired of standin' in one spot and moves around a bit. But wherever you might see him, he is sure to catch your attention! I drop off a bale of hay to him every so often, to keep him from gettin' too hungry. Oh, I forgot to mention, Lon made him out of a bunch of barbed wire! He did a great job of "elk building". That elk has a beautiful set of antlers. The sight of them is enough to make a real live elk sigh with envy. I'm sure many tourists have sneaked closer for a picture.

Lon's elk

PARTING THOUGHTS

I hope that as you rode down the trail of memories with me, that you enjoyed the ride as much as I have. All of the stories in this book are true, but if your recollection of the events differ from mine, well, maybe its because of all the times some critter either kicked me, rolled on me or bucked me off! The people are all real people, my family, neighbors, friends and acquaintances. Hopefully, I have not offended anyone with my brand of humor. If so, it certainly was not my intent.

One of my greatest joys in life has been the privilege of working side by side with a lot of different people. I learned from all of 'em. Getting to know folks and experiencing life situations with them has been my pleasure and a real education. I feel fortunate to have crossed paths with so many great people. If you're in this part of Big Sky Country, drop by, and we'll visit and swap a few more tales. I'd like that!

Claudie Demarest

Gate to Murray Place, Demarest Ranch May 2001

Index

N

O

P

R

S

T

U

W

Y

ORDER FORM

Cardboard Ponies to Buckin' Broncs

YES, I want ___ copy[ies] of *Cardboard Ponies to Buckin' Broncs* at $16.00 (US funds only) each plus $3.00 shipping & handling per 1-2 books to same USA address.

☐ My check or money order payable to Claudie Demarest for $________ is enclosed.

NAME ______________________________

ADDRESS ______________________________

CITY/STATE/ZIP ______________________________

PHONE/EMAIL (optional) ______________________________

Please return to: Claudie Demarest, Box 7, Whitlash, MT 59545